Uncovering Mathematics

with Manipulatives and Calculators

Level 1

Developed by
Jane F. Schielack and Dinah Chancellor

Design by
Reecie Ross

With contributions by
Yolanda Andrade, Bob Fedorisko, Gay Riley-Pfund,
Jan Stevens, Lynn Tanner, and Dianna Tidwell

D1444307

Calculators in Elementary Mathematics

Twenty years ago, one educational question being asked about technology was, "**Should** calculators be used in mathematics in the elementary grades?" The availability of technology in today's world has moved us forward to the question, "**How** can calculators best be used in teaching and learning elementary mathematics?"

The activities in this set of books are examples of situations in which students develop healthy attitudes toward the use of technology, not as a replacement for their mathematical thinking, but as an extension of their mathematical power.

The Level 1 activities, designed for primary grades, build connections between physical representations and the mathematical symbolism of the calculator.

The Level 2 & 3 activities (available separately), designed for intermediate grades, use the calculator as a data-generating device to support exploration of mathematical ideas.

With the calculator, students can explore a problem situation by producing a fairly large set of data in a relatively short period of time. By controlling variables in the problem while collecting the data and organizing the collected data in various ways, students can look for patterns and make conjectures. These patterns and conjectures lead not only to solutions for the given problem, but also to general understanding of important mathematical concepts.

Components of the Activities

Each activity consists of *Teacher Pages* that give directions for the activity and a student *Recording Sheet* that provides a method for organizing collected data.

The *Teacher Pages* for each activity contain:

- The content strand with which the activity is most closely associated: Number Sense; Patterns, Relations, and Functions; Measurement and Geometry; or Probability and Statistics.

- An icon indicating the suggested level of the activity:

  Level 1 Activity

  Level 2 Activity

  Level 3 Activity

  Level 2 and level 3 activities can be found in *Uncovering Mathematics with Manipulatives and Calculators Levels 2 & 3.*

- An **Overview** of the mathematical purpose of the activity.

- A list of related **Mathematical Concepts** addressed in the activity.

- An **Introduction** section, which includes problems that are posed in motivating, real-life settings and often contain literature connections.

- A section for **Collecting and Organizing Data** pertinent to the problem, including questions that require students to think carefully about any models, symbols, or procedures that they have chosen to use.

- A section on **Analyzing Data and Drawing Conclusions**, including questions that require students to look carefully for patterns in the data and translate those patterns into conjectures that can be investigated with examples or verified with mathematical reasoning.

- A concluding section on **Continuing the Investigation** with ideas for related problems and activities.

The student *Recording Sheet* for each activity is designed to provide a slightly organized, but open-ended, structure in which students can successfully record and analyze the data that they collect in the activity. The *Recording Sheets* are also intended to be used by students as models for organizing data in problems encountered outside these activities.

© 1995 Texas Instruments Incorporated. ™ Trademark of Texas Instruments Incorporated.

Goals for Calculator Use in a Mathematics Activity

The design for this collection of activities for integrating the calculator into elementary mathematics instruction is based upon three critical goals.

Goal 1: The activities will address the major components of the elementary mathematics curriculum.

The current recommendations for the elementary mathematics curriculum put forward by the National Council of Teachers of Mathematics states that elementary mathematics instruction should emphasize *problem solving*, *reasoning*, *communication*, and *connections* while engaging students in learning mathematical concepts.

The activities in this collection are clustered into four content strands:

- Number Sense

- Patterns, Relations, and Functions

- Measurement and Geometry

- Probability and Statistics

These four strands are woven together with the integrated use of operations and graphing. Each activity is based on a problem—sometimes real-life, sometimes mathematical, in origin. Each activity is developed around questions that encourage students to analyze data and make inferences, and every activity requires some kind of mathematical communication, either spoken or written.

There are no activities whose *main* goal is to teach a student how to use a particular feature of the calculator. For more information on the specific techniques of using the calculator, refer to the instructional materials for each of the calculators: TI-108, Math Mate™, and the Math Explorer™. You may order the materials by calling **1-800-TI-CARES.**

Goals for Calculator Use in a Mathematics Activity (continued)

Goal 2: The calculator will be an instructional component of activities in which students are engaged in developing understanding of important mathematical ideas.

The use of a calculator in Kindergarten through Grade 6 mathematics is often thought of as a "reward" for completing paper-and-pencil computation or as "support" for a student having difficulty with paper-and-pencil computation. The use of the calculator in these activities, however, is only one component of an instructional plan for teaching and learning mathematics.

In these activities, the calculator is used mainly for generating data, with students deciding what data is needed and what processes should be used for generating it. Students must make decisions about what should and what should not be changed in the problem situation as the data is generated and how the data should be organized in order to discover the patterns that lead to important mathematical ideas.

On the left side of the *Teacher Pages* are questions that address the content of the activity and may or may not pertain to the use of the calculator. Along the right side of the *Teacher Pages*, marked with the calculator icon are questions that particularly address the role of the calculator in the activity.

Goal 3: The calculator will be used in an environment that encourages reasoning and communication.

Most of the activities in this collection are designed for students working together in small groups of two, three, or four. To encourage interaction within groups, the students in each group can work together using one copy of the *Recording Sheet*. Each member of a group can take on a particular role in collecting or organizing the data, and these roles can be rotated so that each member is involved in each aspect of the problem. With designated roles in solving the problem, a group can work with a single calculator, rotating its use among the members.

The purpose of the calculator in most of the activities is to generate questions rather than answers. The solutions to the problems and the conjectures about abstract mathematical ideas develop from students and teachers analyzing the data and discussing the questions that are generated in the activity.

Table of Contents

Level 1

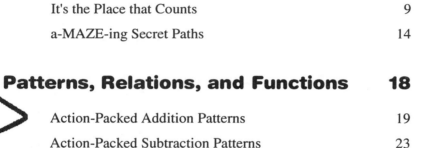

Level 1

Activities:

1 Action-Packed Stories

2 Patterns in Counting

3 It's the Place that Counts

4 a-MAZE-ing Secret Paths

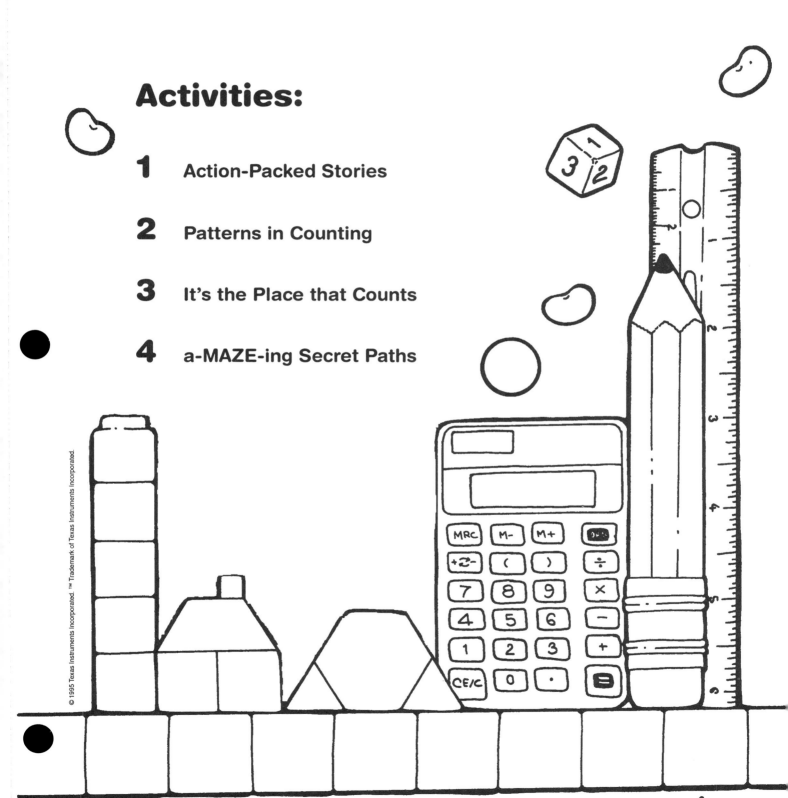

Action-Packed Stories

Math Concepts

- whole numbers
- addition
- subtraction

Materials

- TI-108, Math Mate™, Math Explorer™
- counters
- **Action-Packed Stories** recording sheets
- pencils

Overview

Students will choose a number, create stories to go with the number, illustrate their stories, and represent the action in their stories on the calculator.

Introduction

1. Have students choose a number. Write it in the center of a recording sheet transparency.

2. Work with the whole class to make up a story using that number.

   Example: Here is a story for the number 5. Three bears were asleep in their cave. Two more bears wandered in, and then there were five bears in the cave.

3. Model the action in the story on the overhead projector using counters, beans, or a transparency of the bear pictures provided on page 5. You could also ask a student to draw a picture of the action in the story.

4. Have students represent the action in the story on their calculators and write the problem in the space provided on their recording sheets.

5. Have students work in pairs to make up three more stories to go with the same number. Then have them draw illustrations and use their calculators to represent the numbers in the new stories.

   To help spark some creative stories, read *Ten Black Dots* by Donald Crews, *One Hungry Monster* by Susan Heyboer O'Keefe, or *Anno's Counting House* by Mitsumasa Anno with students.

6. Have students select another number and repeat the process.

Historias activas

Conceptos matemáticos

- números enteros
- suma
- resta

Materiales

- TI-108, Math Mate™, Math Explorer™
- contadores
- hojas de registro de **Historias activas**
- lápices

Resumen

Los alumnos elegirán un número, crearán historias para ese número, ilustrarán sus historias y representarán la acción en sus calculadoras.

Introducción

1. Los alumnos deberán elegir un número. Escríbalo en el centro de una transparencia de la hoja de registro.

2. Trabaje junto con toda la clase para inventar una historia que pueda aplicarse a dicho número.

    Ejemplo: Esta es una historia para el número 5. Tres osos dormían en su cueva. A ella entraron otros dos osos y entonces había cinco osos en la cueva.

3. Represente la acción de la historia en el proyector de transparencias utilizando contadores, frijoles o una transparencia de dibujos de osos de la página 5. También puede pedir a uno de los alumnos que haga un dibujo de la acción de la historia.

4. Pida a los alumnos que representen la acción de la historia en sus calculadoras y que escriban el problema en el espacio provisto para tal fin en la hoja de registro.

5. En grupos de a dos, los alumnos deberán inventar tres historias más que se apliquen al mismo número. Luego, pídales que hagan ilustraciones y utilicen sus calculadoras para representar los números en las nuevas historias.

    Para lograr que se les ocurran historias creativas, lea a los alumnos *Ten Black Dots* de Donald Crews, *One Hungry Monster* de Susan Heyboer O'Keefe o *Anno's Counting House* de Mitsumasa Anno.

6. Pida a los alumnos que elijan un nuevo número y repitan el proceso.

Action-Packed Stories (continued)

Collecting and Organizing Data

While students are creating situations to illustrate their number stories, ask questions such as:

- How could you show the action in your story with counters? With a picture?

- What math operation could show the same action?

- Can you make up a different story using the same action?

- Can you make up a different action story about the same number?

- Can you make up a story using several different actions? How could you illustrate that?

How can the action in your story be represented on the calculator?

What do the numbers you are pressing on your calculator represent in your story?

Analyzing Data and Drawing Conclusions

After students have made up their stories, illustrated them, and represented them on their calculators, have them work as a whole group to analyze their stories. Ask questions such as:

- Will you share your story and describe the action in it?

- Which operation did you use to represent the action in your story?

How did you use the calculator to help you show the action in your story?

Does the order in which you entered the numbers in your calculator matter to your story? Why or why not?

Continuing the Investigation

Have students:

- Select other numbers and repeat the process for each number.

- Record number sentences to describe the actions in the new stories and decide how to use the calculator to represent the numbers.

Historias activas (continuación)

Cómo reunir y organizar los datos

Mientras los alumnos imaginan situaciones para ilustrar las historias con números, formule preguntas tales como:

- ¿Cómo podrían reproducir la acción de la historia utilizando contadores? ¿Con un dibujo?

- ¿Qué operación matemática podría reproducir la misma acción?

- ¿Pueden inventar otra historia con la misma acción?

- ¿Pueden inventar otra historia con el mismo número?

- ¿Pueden inventar una historia en la que utilicen diferentes acciones? ¿Cómo podrían ilustrarla?

🖩 ¿Cómo se puede representar la acción de su historia en la calculadora?

🖩 ¿Qué parte de su historia representan los números que están oprimiendo en la calculadora?

Cómo analizar los datos y sacar conclusiones

Después de que los alumnos hayan inventado sus historias, las hayan ilustrado y las hayan representado en las calculadoras, trate de que las analicen en conjunto formulando preguntas como:

- ¿Podrían contarnos su historia y describir la acción que representa?

- ¿Qué operación utilizaron para representar la acción de su historia?

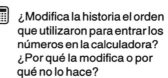

🖩 ¿Cómo utilizaron la calculadora para representar la acción de su historia?

🖩 ¿Modifica la historia el orden que utilizaron para entrar los números en la calculadora? ¿Por qué la modifica o por qué no lo hace?

Cómo continuar la investigación

Pida a sus alumnos que:

- Elijan otros números y repitan el proceso con cada número.

- Registren las operaciones matemáticas que describen las acciones de las nuevas historias y decidan cómo utilizar la calculadora para representar los números.

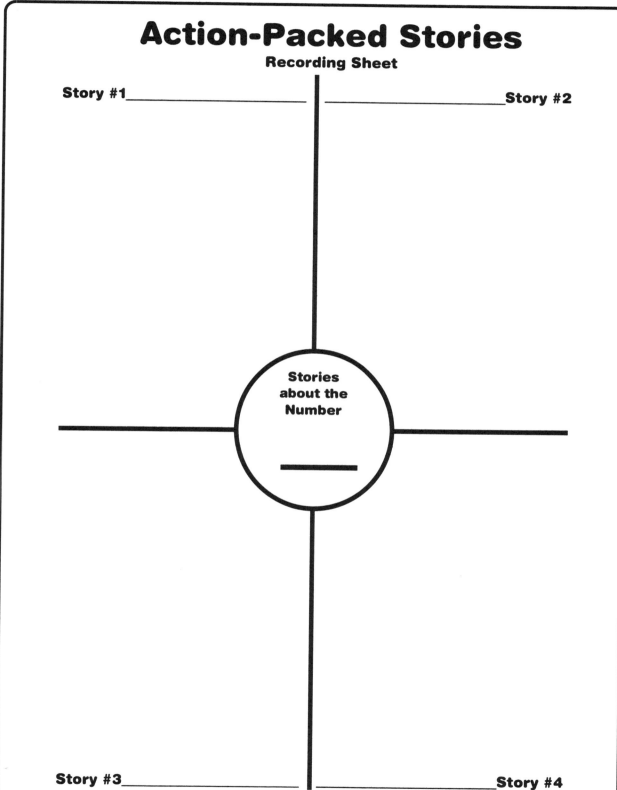

Name:

Action-Packed Stories

Recording Sheet

Story #1 ___ / ___ Story #2

Stories about the Number ___

Story #3 ___ / ___ Story #4

Number Sense

Uncovering Mathematics with Manipulatives and Calculators Level 1

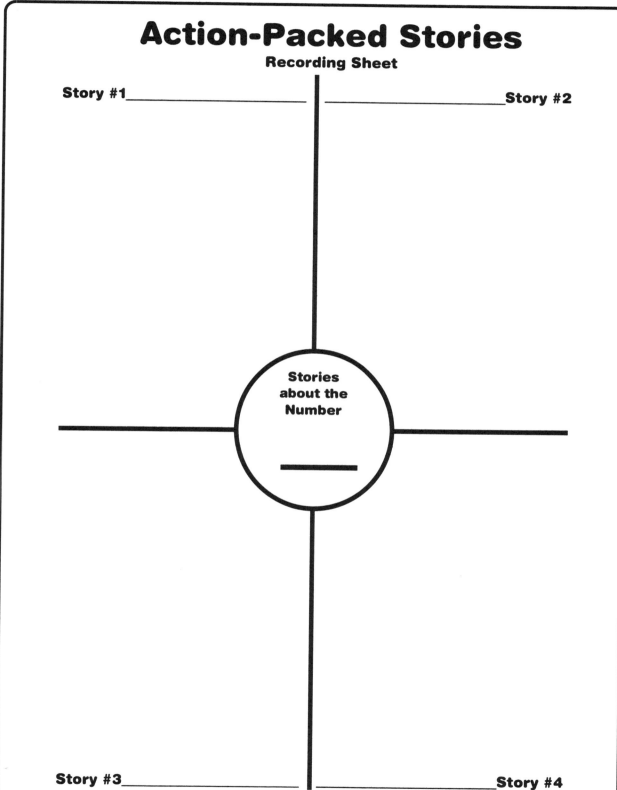

Nombre:

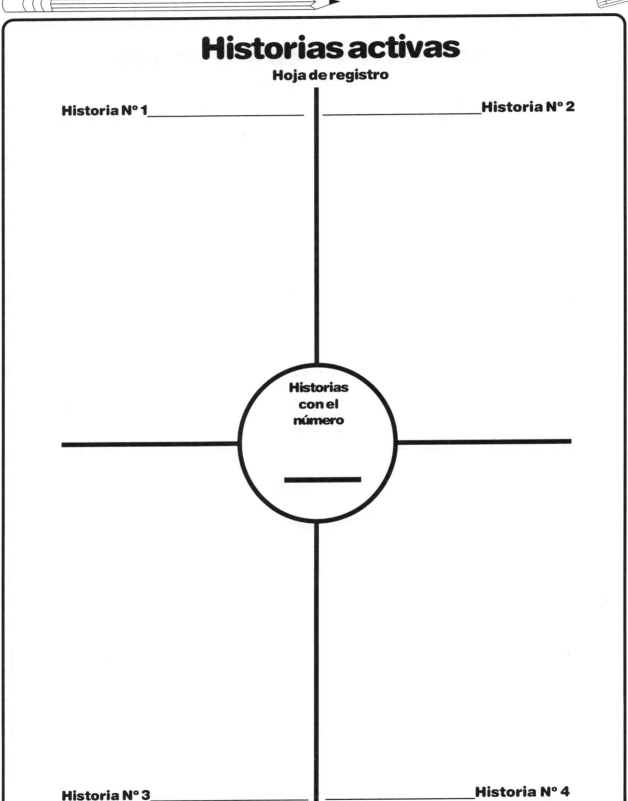

Historias activas
Hoja de registro

Historia N° 1_____   _____Historia N° 2

**Historias
con el
número**

_____

Historia N° 3_____   _____Historia N° 4

Action-Packed Stories
Counting Bears

Historias activas
Osos para contar

Revelación de Matemáticas con contadores móviles y calculadoras - Nivel 1

Patterns in Counting

Math Concepts

- patterns
- whole numbers
- comparing numbers
- ordering numbers
- place value
- addition

Materials

- TI-108, Math Mate™, Math Explorer™
- beans or centimeter cubes
- **Patterns in Counting** recording sheets
- pencils

Overview

Students will use the calculator to count sets of concrete objects, connect number symbols to quantities, and look for patterns in the number symbols.

Introduction

1. Have each student hold ten small beans or centimeter cubes in one hand.

2. Ask students: How many beans (cubes) do you think you could hold in one hand? Why do you think that? Have students record their estimates on their recording sheets.

3. Have each student grab one handful of beans (cubes) from a bag and place them beside a recording sheet.

4. Have students:

   a. Prepare their calculators to count by entering ⊞**1**.

   b. Place one bean (cube) in the top left square of the hundred grid on their recording sheets.

   c. Press ⊟ to display **1**.

   d. Place a second bean (cube) in the next square across on the grid.

   e. Press ⊟ to display **2**.

5. Have students continue counting their beans (cubes) by placing them one at a time on the grid and pressing ⊟ to display the symbol on the calculator for the number of beans.

   Note: Students can mark (color) the squares as they place each bean (cube) on the grid so that they can see the pattern when they remove the beans (cubes).

6. Read *1 Hunter* by Pat Hutchins, *One Gorilla* by Atsuko Morozumi, or *Rooster's Off to See the World* by Eric Carle to students to reinforce the pattern of "one more."

Patrones para contar

Conceptos matemáticos

- patrones
- números enteros
- comparación numérica
- orden de los números
- valor de posición
- suma

Materiales

- TI-108, Math Mate™, Math Explorer™
- frijoles o cubos de 1 cm de lado
- hojas de registro de **Patrones para contar**
- lápices

Resumen

Los alumnos utilizarán la calculadora para contar conjuntos de objetos reales, relacionar los símbolos numéricos con las cantidades y buscar patrones que se apliquen a los símbolos numéricos.

Introducción

1. Pida a cada alumno que tome diez frijoles pequeños o cubos de 1 cm de lado en una mano.

2. Pregunte: ¿Cuántos frijoles (cubos) piensan ustedes que pueden sostener en una mano? ¿Por qué? Cada alumno deberá anotar su estimación en las hojas de registro.

3. Cada alumno deberá extraer de una bolsa un puñado de frijoles (cubos) y colocarlo junto a la hoja de registro.

4. Pida a los alumnos que:

    a. Entren ⊞ **1** en la calculadora para prepararla para contar.

    b. Coloquen uno de sus frijoles (cubos) en el casillero izquierdo superior de la cuadrícula de centena de la hoja de registro.

    c. Opriman la tecla ⊟ para que aparezca el **1** en el visor de la calculadora.

    d. Coloquen un segundo frijol (cubo) en el siguiente casillero de la cuadrícula.

    e. Opriman ⊟ para que aparezca el **2**.

5. Los alumnos deberán continuar con la cuenta de frijoles (cubos), colocándolos de a uno en la cuadrícula para luego presionar ⊟ y lograr que aparezca en el visor de la calculadora el símbolo numérico que equivalga a la cantidad de frijoles.

    Nota: Los alumnos pueden marcar (colorear) los casilleros a medida que colocan cada frijol (cubo) en la cuadrícula, a modo de ver el patrón cuando quiten los frijoles (cubos).

6. Lea a los alumnos *I Hunter* de Pat Hutchins, *One Gorilla* de Atsuko Morozumi o *Rooster's Off to See the World* de Eric Carle para reforzar el patrón de "uno más".

Patterns in Counting (continued)

Collecting and Organizing Data

While students explore with the beans (cubes) and calculators, ask questions such as:

- How many beans (cubes) fit across the grid?

- What patterns do you notice in the numbers on the calculator as you fill up the grid?

- When do the numbers start using two spaces on the calculator?

- Which part of the number changes as you add each bean (cube)?

🖩 After you press ⊟, what does the number on the calculator display show you?

🖩 Why do you think you enter ⊞1 to prepare the calculator to count?

🖩 What do you think would happen if you entered ⊞ 2 instead of ⊞1 to begin?

Analyzing Data and Drawing Conclusions

After students have counted their different groups of beans (cubes), have them work as a whole group to analyze their observations. Ask questions such as:

- What patterns did you notice in the numbers while you were counting?

- How are the beans (cubes) on the grid and the numbers on the calculator connected?

- How many different ways can you describe the number of beans (cubes) you were able to hold in your hand?

- Who was able to hold the greatest number of beans (cubes)? How do we know?

- Who grabbed the smallest number of beans (cubes)? How do we know?

- Why did we end up with different numbers of beans?

🖩 How did you use the calculator to help you count?

🖩 How could you use the calculator to count two beans (cubes) at one time? Three? More than three?

Continuing the Investigation

- How many beans (cubes) do you think you could hold in two hands? Why? Use the calculator and the hundred grid to test your conjecture.

- If you had a partner who was also holding beans (cubes) in both hands, how many could you hold together? Why? Use the calculator and the hundred grid to test your conjecture.

Patrones para contar (continuación)

Cómo reunir y organizar los datos

Mientras los alumnos investigan la relación entre los frijoles (cubos) y la calculadora, formule preguntas tales como:

- ¿Cuántos frijoles (cubos) caben en horizontal en la cuadrícula?

- ¿Cuáles son los patrones que advierten en los números que aparecen en la calculadora a medida que se va llenando la cuadrícula?

- ¿Cuándo comienzan los números a ocupar dos espacios en la calculadora?

- ¿Qué parte del número se modifica cada vez que se agrega un frijol (cubo)?

⊞ Después de oprimir ⊟ , ¿qué muestran los números que aparecen en el visor de la calculadora?

⊞ ¿Por qué tienen que entrar ⊞1 para preparar la calculadora para contar?

⊞ ¿Qué piensan que sucedería si entraran ⊞ 2 en vez de ⊞ 1 para comenzar?

Cómo analizar los datos y sacar conclusiones

Cuando los alumnos hayan terminado de contar los diferentes grupos de frijoles (cubos), pídales que trabajen en conjunto para analizar sus observaciones y formúleles las siguientes preguntas:

- ¿Qué patrones advirtieron en los números mientras contaban?

- ¿Cómo se relacionan los frijoles (cubos) de la cuadrícula con los números de la calculadora?

- ¿Cuántas maneras distintas se les ocurren para describir la cantidad de frijoles (cubos) que pudieron sujetar en la mano?

- ¿Quién pudo sujetar la mayor cantidad de frijoles (cubos)? ¿Cómo obtenemos ese dato?

- ¿Quién pudo sujetar la menor cantidad de frijoles (cubos)? ¿Cómo obtenemos ese dato?

- ¿Por qué todos sujetamos diferente cantidad de frijoles?

⊞ ¿Cómo utilizaron la calculadora para contar?

⊞ ¿Cómo podrían utilizar la calculadora para contar de a dos frijoles (cubos) por vez? ¿Tres? ¿Más de tres?

Cómo continuar la investigación

- ¿Cuántos frijoles (cubos) piensan que podrían sujetar en ambas manos? ¿Por qué? Utilicen la calculadora y la cuadrícula de centena para comprobar sus conjeturas.

- Si su compañero también sujeta frijoles (cubos) en ambas manos, ¿cuántos podrían sujetar en conjunto? ¿Por qué piensan eso? Utilicen la calculadora y la cuadrícula de centena para comprobar sus conjeturas.

It's the Place that Counts (continued)

Introduction (continued)

8. Have students work in pairs. As one partner puts some place-value pieces in the "Base Ten Pieces" space on the recording sheet, the other partner enters the appropriate commands in the calculator to count the pieces.

9. Have students draw pictures of the base ten pieces they chose. Or have them make a picture using the base ten pieces provided on page 13. Ask students to describe how they used the calculator.

Collecting and Organizing Data

While students explore with the place-value materials and their calculators, ask questions such as:

- How much is this piece worth? How do you know?

- What patterns do you notice when you are counting the tens?

- What do the numbers you are reading on the calculator represent?

- Do you always have to do the tens first? What would happen if you did the ones first?

⊞ Why do you enter ⊞ **10** to begin counting the tens?

⊞ Why do you change to ⊞ **1** when you are counting the ones?

⊞ What is happening when you press ⊟ ?

Analyzing Data and Drawing Conclusions

After students have counted several different groups of place-value materials, have them work as a whole group to analyze their observations. Ask questions such as:

- What patterns did you notice in the numbers while you were counting?

- How are the tens and ones pieces you chose and the numbers on the calculator connected?

⊞ How did you use the calculator to represent the total?

⊞ How could you use the calculator to count hundreds? Thousands?

La posición es lo que cuenta (continuación)

Introducción (continuación)

8. Agrupe a los alumnos por pares. A medida que un alumno coloca algunas piezas de valor de posición en el espacio "Unidades en numeración decimal" de la hoja de registro, su compañero entrará en la calculadora los números y símbolos que correspondan para contar las piezas.

9. Pida a los alumnos que dibujen las unidades en numeración decimal que eligieron. También podrán hacer un dibujo utilizando las unidades en numeración decimal provistas en la página 13. Trate de que los alumnos describan la forma en que utilizaron la calculadora.

Cómo reunir y organizar los datos

Mientras los alumnos investigan con los materiales de valor de posición y sus calculadoras, formule preguntas tales como:

- ¿Cuánto vale esta tira? ¿Cómo lo saben?

- ¿Qué patrones advierten cuando cuentan las decenas?

- ¿Qué representan los números que están leyendo en la calculadora?

- ¿Se debe comenzar siempre por las decenas? ¿Qué sucedería si comenzaran con las unidades?

¿Por qué deben entrar + 10 para comenzar a contar las decenas?

¿Por qué deben cambiar a + 1 cuando cuentan las unidades?

¿Qué sucede cuando oprimen = ?

Cómo analizar los datos y sacar conclusiones

Después de que los alumnos hayan contado diferentes grupos de materiales de valor de posición, trate de que en conjunto analicen sus observaciones y para ello formule preguntas tales como:

- ¿Qué patrones advirtieron en los números mientras contaban?

- ¿Cómo se conectan las decenas y las unidades que eligieron con los números que aparecen en la calculadora?

¿Cómo utilizazaron la calculadora para representar el total?

¿Cómo podrían utilizar la calculadora para contar centenas? ¿Unidades de mil?

It's the Place that Counts (continued)

Continuing the Investigation

Have students:

- Investigate the question: How many tens does it take to reach one hundred? Use place-value materials and the calculator to display some numbers that include hundreds.

- Investigate the question: Do you always have to count the largest pieces first? What would happen if you counted the ones, then the tens, and then the hundreds? Try it. Record what the calculator displays and compare it to counting the hundreds first, then the tens, and then the ones.

La posición es lo que cuenta (continuación)

Cómo continuar la investigación

Indique a sus alumnos que:

- Investiguen lo siguiente: ¿Cuántas decenas necesito para llegar a cien? Utilicen los materiales de valor de posición y la calculadora para mostrar algunos números que tengan centenas.

- Investiguen lo siguiente: ¿Debes siempre comenzar a contar por las piezas mayores? ¿Qué sucedería si comenzaran por las unidades, siguieran por las decenas y terminaran por las centenas? Inténtenlo. Registren lo que aparece en el visor de la calculadora y compárenlo con el resultado de contar primero las centenas, después las decenas y por último las unidades.

Name:

It's the Place that Counts
Recording Sheet

Collecting and Organizing Data

Base Ten Pieces	Base Ten Pieces

Number: _____   Number: _____

How I used the calculator:

Questions we thought of while we were doing this activity:

Nombre:

La posición es lo que cuenta
Hoja de registro

Cómo reunir y organizar los datos

Unidades en numeración decimal	Unidades en numeración decimal

Número: _____

Número: _____

De qué manera utilicé la calculadora:

Algunas preguntas que surgieron mientras realizábamos esta actividad:

It's the Place that Counts

Ones squares　　　　　　　　　　　　　　**Tens sticks**

Hundred square　　　　　　　　　　　　　**Tens sticks**

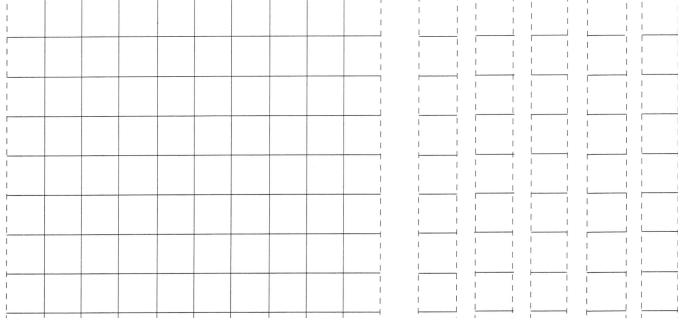

La posición es lo que cuenta

Casilleros de unidades

Tiras de decenas

Cuadrado de centena

Tiras de decenas

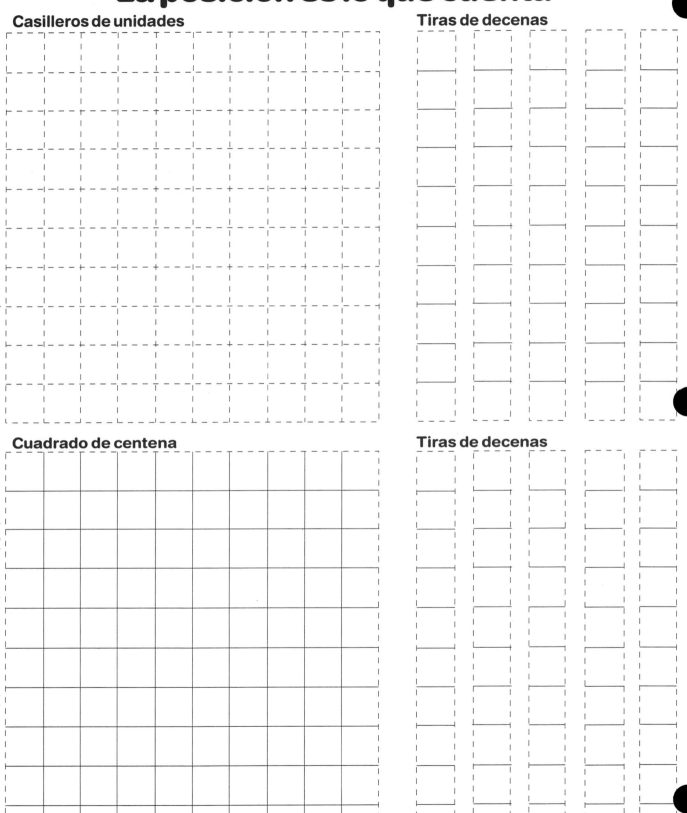

a-MAZE-ing Secret Paths

Math Concepts

- whole numbers
- place value
- comparing numbers
- estimation
- addition
- subtraction

Materials

- TI-108, Math Mate™, Math Explorer™
- **a-MAZE-ing Secret Paths** recording sheets
- pencils

Overview

Students will investigate patterns in place value and the effects of adding and subtracting by using the calculator to explore relationships between numbers on the hundred chart.

Introduction

1. Have students choose a number from the hundred chart on their recording sheets. Then have them look at the number to the right of it. Ask them how this number is different from the first number.

   Example: 28 is "one" more than 27.

2. Have each student enter the number he or she chose in the calculator. Ask: What would you do to change the calculator display so that it shows the value of the number in the square to the right?

   Example: Enter **27** ⊞ **1** ⊟ to change 27 to 28.

3. Have each student investigate the number to the left, below, and above his or her original number in the same way.

   After each investigation, ask: What did you have to do on your calculator to change the number in the display to the number in the square to the left , below, and above the number you chose?

4. Model how to play **a-MAZE-ing Secret Paths** for students using a transparency of the recording sheet. Draw a path on the hundred chart and record the path with numbers and operations. Next, have students follow the path using a calculator.

   Example: For a beginning number of 76 and an ending number of 64, you could record this path: 76 - 1 - 1 - 10. Read the path to students and have them follow the steps on their calculators. Students should arrive at your ending number of 64. (For this example, students should press **76** ⊟ **1** ⊟ **1** ⊟ **10** ⊟.)

Sorprendente laberinto secreto

Conceptos matemáticos

- números enteros
- valor de posición
- comparación numérica
- estimación
- suma
- resta

Materiales

- TI-108, Math Mate™, Math Explorer™
- hojas de registro de **Sorprendente laberinto secreto**
- lápices

Resumen

Los alumnos investigarán los patrones en el valor de posición y los efectos de las operaciones de suma y resta. Usarán la calculadora para explorar las relaciones entre los números de la tabla de centenas.

Introducción

1. Pida a los alumnos que elijan un número de la tabla de centenas en la hoja de registro. Luego, pídales que miren el número que se encuentra a la derecha del número elegido. Pregunte en qué se diferencia éste del primer número.

   Ejemplo: 28 es "uno" más que 27.

2. Pida a sus alumnos que entren el número elegido en la calculadora. Pregunte: ¿Qué podrían hacer para que en el visor de la calculadora aparezca el valor del número situado en el casillero de la derecha?

   Ejemplo: Entraría **27** ⊞ **1** ⊟ para pasar de 27 a 28.

3. Pida a cada alumno que investigue de la misma manera el número situado a la izquierda, el número debajo y el número arriba de su número inicial.

   A medida que terminen cada paso, pregunte: ¿Qué tuvieron que hacer en la calculadora para hacer coincidir el número que aparecía en el visor con el número elegido en el casillero de la izquierda, debajo o arriba)?

4. Utilice una transparencia de la hoja de registro para explicar a los alumnos el juego del **Sorprendente laberinto secreto**. Dibuje un sendero en la tabla de la centena y regístrelo con números y operaciones. A continuación, pida a los alumnos que sigan el sendero utilizando una calculadora.

   Ejemplo: Para un número inicial 76 y un número final 64, usted puede registrar el siguiente sendero: 76 - 1 - 1 - 10. Lea el sendero a los alumnos y pídales que sigan los pasos en sus calculadoras. Los alumnos deben alcanzar el número final, 64. (Para este ejemplo, los alumnos deben oprimir **76** ⊟ **1** ⊟ **1** ⊟ **10** ⊟ .)

a-MAZE-ing Secret Paths (continued)

Introduction (continued)

5. Have students choose two numbers on the hundred chart and secretly draw a path between those two numbers. The path must move from square to square either horizontally or vertically, **not diagonally**. Using numbers and operation symbols, have students individually record the steps in their paths on the stone path provided on their recording sheets.

6. Now have students work in pairs. Tell the first partner to read the beginning number and each step for moving through the path to reach the ending number (without revealing the ending number). Have the second partner listen, use a calculator to enter each number and operation, and press $=$ at the end of the directions to display the ending number. Then have students compare the recorded ending number with the ending number displayed on the calculator.

7. Have the partners switch roles and play the **a-MAZE-ing Secret Paths** game again.

Collecting and Organizing Data

While students explore their mazes, ask questions such as:

- Which steps in your path move to the left? How far do they move to the left? Which steps in your path move to the right? How far do they move to the right? How about up and down?

- What happens to the numbers when your path moves up and down?

- What happens to the numbers when your path moves to the left or right?

- How do you record this in your description of your path?

- How can you make your path begin and end at the same number (make a loop)?

- How can you make certain geometric shapes with your path?

How can you use the calculator to record the steps in the path that move to the right? To the left? Up? Down?

How are the numbers you are seeing on the calculator connected to the numbers on the hundred chart?

© 1995 Texas Instruments Incorporated. ™ Trademark of Texas Instruments Incorporated.

Sorprendente laberinto secreto (continuación)

Introducción (continuación)

5. Indique a los alumnos que elijan dos números de la tabla de la centena y que, secretamente, tracen un sendero entre estos dos números. El sendero debe avanzar de casillero en casillero en sentido horizontal o vertical, **nunca en forma diagonal**. Pida a los alumnos que utilicen números y símbolos de operaciones en el sendero de piedras que hay en su hoja de registro para anotar individualmente los pasos de su sendero.

6. Agrupe a sus alumnos en pares. Indique a uno de ellos que le diga a su compañero su número inicial y los pasos que deberá seguir para desplazarse por el sendero y llegar al número final (sin revelarle el número final). Pida al otro niño que escuche y utilice su calculadora para entrar cada número y operación, y oprima ⊟ al finalizar las instrucciones para visualizar el número final. Solicite a sus alumnos que comparen el número final registrado con el número final que aparece en el visor de la calculadora.

7. Solicite a los alumnos que cambien funciones y jueguen nuevamente al **Sorprendente laberinto secreto**.

Cómo reunir y organizar los datos

Mientras los alumnos exploran sus laberintos, formule preguntas tales como:

*   ¿Cuáles son los pasos del sendero para desplazarse hacia la izquierda? ¿Cuán lejos se desplazan hacia la izquierda? ¿Cuáles son los pasos para desplazarse hacia la derecha? ¿Cuán lejos se desplazan hacia la derecha? ¿Y hacia arriba o hacia abajo?

*   ¿Qué sucede con los números cuando sus senderos avanzan hacia arriba y hacia abajo?

*   ¿Qué sucede con los números cuando sus senderos avanzan hacia la izquierda o hacia la derecha?

*   ¿Cómo registraron estas acciones en las descripciones de sus senderos?

*   ¿Cómo lograrán que sus senderos empiecen y terminen en un mismo número (haga un lazo)?

*   ¿Cómo se pueden realizar determinadas formas geométricas con el sendero?

🖩 ¿Cómo pueden utilizar la calculadora para registrar los pasos del sendero que hacen desplazar hacia la derecha? ¿Hacia la izquierda? ¿Hacia arriba? ¿Hacia abajo?

🖩 ¿Cómo se relacionan los números que aparecen en el visor de la calculadora con los números de la tabla de la centena?

a-MAZE-ing Secret Paths (continued)

Analyzing Data and Drawing Conclusions

After students have taken turns exploring several paths, have them work as a whole group to analyze the descriptions of their paths. Ask questions such as:

- What information did you use to decide how to record your path?

- Did your directions lead your partner to display the correct ending number on the calculator? Why or why not?

- How could you change your description and still lead your partner to the correct ending number on the same path?

- What happens if you change the order of the steps in your description? Does it change your path? Does it change your ending number?

- Could you make your path shorter, easier to follow? How?

If your partner displayed the wrong ending number, how could you use that number to decide how to change the description of your path?

Continuing the Investigation

Have students:

- Draw three more paths between the same beginning and ending numbers and compare their descriptions of the paths. Ask: How are they alike? How are they different? Can you find anything in common among all four paths?

- Investigate the questions: What if your beginning and ending numbers were equal to each other (a loop)? How would this affect the kinds of paths you could draw? How would this affect the descriptions of your paths?

- Investigate the question: If the rules changed to allow diagonal steps in your path, how would this affect the description of your path?

Sorprendente laberinto secreto (continuación)

Cómo analizar los datos y sacar conclusiones

Una vez que los alumnos se hayan turnado para explorar varios senderos, trate de que en conjunto analicen las descripciones de sus senderos, y para ayudarlos formúleles preguntas tales como:

- ¿Qué información utilizaron para decidir la forma de registrar sus senderos?

- ¿Le sirvieron al compañero sus indicaciones para llegar al número final correcto en la calculadora? ¿Por qué o por qué no?

- ¿De qué manera podrían modificar la descripción para que su compañero llegue al número final correcto en el mismo sendero?

- ¿Qué sucede si modifican el orden de los pasos de su descripción? ¿Se modifica el sendero? ¿Se modifica el número final?

- ¿Podrían hacer su sendero más breve, más fácil de recorrer? ¿Cómo?

Si su compañero llegó a un número final incorrecto, ¿de qué manera podrías utilizar ese número para decidir como modificar la descripción de tu sendero?

Cómo continuar la investigación

Indique a sus alumnos que:

- Dibujen otros tres senderos entre los mismos números inicial y final. Pídales que comparen las descripciones de sus senderos. ¿En qué se parecen? ¿En qué se diferencian? ¿Tienen los cuatro senderos aspectos en común?

- Investiguen lo siguiente: ¿Qué sucede si el número inicial y el número final fueran iguales (un lazo)? ¿De qué manera afectaría esta situación a los tipos de senderos que podrías dibujar? ¿Y de qué manera afectaría la descripción de los senderos?

- Investiguen lo siguiente: Si cambiáramos las reglas del juego y se pudieran dar pasos en diagonal, ¿de qué manera afectaría esta situación la descripción del sendero?

Name:

a-MAZE-ing Secret Paths

Recording Sheet

Collecting and Organizing Data

Hundred Chart

1	2	3	4	5	6	7	8	9	10
11	12	13	14	15	16	17	18	19	20
21	22	23	24	25	26	27	28	29	30
31	32	33	34	35	36	37	38	39	40
41	42	43	44	45	46	47	48	49	50
51	52	53	54	55	56	57	58	59	60
61	62	63	64	65	66	67	68	69	70
71	72	73	74	75	76	77	78	79	80
81	82	83	84	85	86	87	88	89	90
91	92	93	94	95	96	97	98	99	100

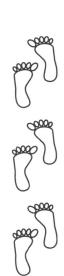

Move from my beginning number to my ending number by following this path.

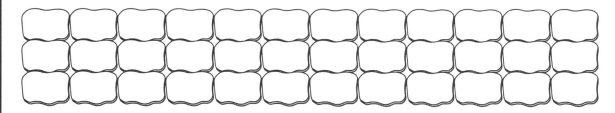

Patterns we found while we were doing this activity:

Nombre:

Sorprendente laberinto secreto

Hoja de registro

Cómo reunir y organizar los datos

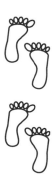

Tabla de la centena

1	2	3	4	5	6	7	8	9	10
11	12	13	14	15	16	17	18	19	20
21	22	23	24	25	26	27	28	29	30
31	32	33	34	35	36	37	38	39	40
41	42	43	44	45	46	47	48	49	50
51	52	53	54	55	56	57	58	59	60
61	62	63	64	65	66	67	68	69	70
71	72	73	74	75	76	77	78	79	80
81	82	83	84	85	86	87	88	89	90
91	92	93	94	95	96	97	98	99	100

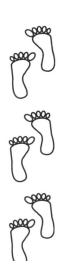

Puedes llegar desde mi número inicial a mi número final si sigues este sendero.

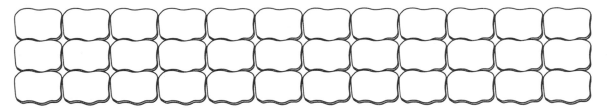

Patrones que descubrimos mientras realizábamos esta actividad:

Level 1

Activities:

1 Action-Packed Addition Patterns

2 Action-Packed Substraction Patterns

3 Predictable Patterns with Addition

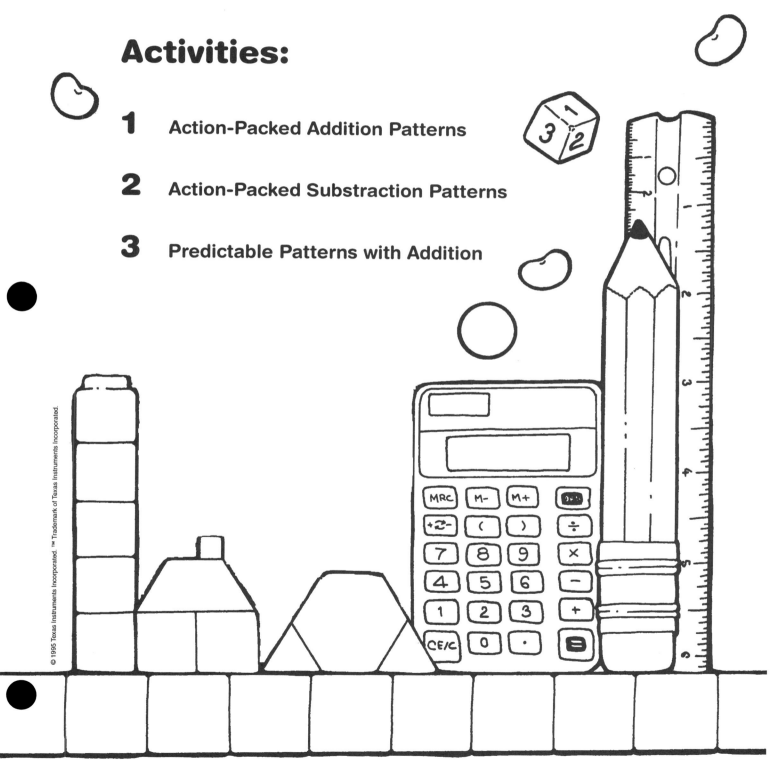

Action-Packed Addition Patterns

Math Concepts	Overview
• whole numbers • addition • patterns **Materials** • TI-108, Math Mate™, Math Explorer™ • manipulatives or counters • **Action-Packed Addition Patterns** recording sheets • pencils	Students will use manipulatives and calculators to explore what happens when they change one number at a time in an addition number sentence. They will also record and describe the pattern that develops.

Introduction

> The **Action-Packed Stories** activity on page 2 should be completed before beginning this activity.

1. Have each student choose one of his or her **Action-Packed Stories** and write the number sentence that goes with it on the recording sheet: 2 + 1 = 3, for example. Next, ask them to model the story they chose with manipulatives.

   Or, have students read *There Was an Old Lady Who Swallowed a Fly* by Pam Adams and write number sentences for the story. Then ask them to choose one of those number sentences and write it on their recording sheets. Next, ask them to model the story they chose with manipulatives.

2. Ask students to choose one number in the story to change. Have them circle the number in the number sentence on their recording sheets.

   Examples:
   In the number sentence 2 + 1 = 3, change 1 to 2; the new number sentence is 2 + 2 = 4.

   In *There Was an Old Lady Who Swallowed a Fly*, 1 lady + 1 animal = 2 things, 1 lady + 2 animals = 3 things, etc.

3. Now have students model the new number sentence with manipulatives, and discuss how the story should change to work with the new number sentence.

4. Have students continue to change the same number, tell the new story it represents, and model it with manipulatives.

Patrones activos de la suma

Conceptos matemáticos	**Resumen**
• números enteros • suma • patrones	Los alumnos utilizarán contadores móviles y calculadoras para explorar lo que sucede cuando modifican los números de una suma de a uno por vez. También registrarán y describirán el patrón que se desarrolla.
Materiales	
• TI-108, Math Mate™, Math Explorer™ • contadores móviles o contadores • hojas de registro de **Patrones activos de la suma** • lápices	

Introducción

> Antes de iniciar esta actividad se deberá haber completado **Historias activas** de la página 2.

1. Pida a los alumnos que elijan alguna de sus **Historias activas** y que escriban la operación que corresponde a ella en la hoja de registro: por ejemplo, 2 + 1 = 3. A continuación, pida a sus alumnos que representen con los contadores móviles la historia elegida.

   También, los alumnos pueden leer *There Was an Old Lady Who Swallowed a Fly* de Pam Adams y escribir operaciones para la historia. Luego, pídales que elijan una de las operaciones de la historia y que la escriban en la hoja de registro. A continuación, los niños deberán representar la historia con sus contadores móviles.

2. Pida a los alumnos que elijan un número de la historia que desean modificar. Indíqueles que dibujen un círculo alrededor del número de la operación que anotaron en la hoja de registro.

   Ejemplos:
   En la operación 2 + 1 = 3, modifique el 1 por 2; la nueva operación es 2 + 2 = 4.

   En el cuento *There Was an Old Lady Who Swallowed a Fly*, 1 señora + 1 animal = 2 objetos, 1 señora + 2 animales = 3 objetos, etc.

3. A continuación, pida a los alumnos que representen la nueva operación con contadores móviles y debatan la manera en la que la historia debería modificarse para que corresponda a la nueva operación.

4. Pida a los alumnos que continúen modificando el mismo número, inventen la nueva historia que representa y la representen usando los contadores móviles.

Action-Packed Addition
Patterns (continued)

Introduction (continued)

5. Record all of the changes in the number sentences on the recording sheets and discuss the patterns that develop.

   Example:
   2 + 1 = 3
   2 + 2 = 4
   2 + 3 = 5
   2 + 4 = 6
   2 + 5 = 7

6. Have students choose a similar situation using larger numbers. Then have them use their calculators to find the number sentence patterns and write them on their recording sheets.

Collecting and Organizing Data

While students are exploring their patterns, ask questions such as:

- What happened to the sum each time you changed an addend? Did it get smaller or larger? Why?

- Could you show me with your manipulatives? Would your story stay the same? How might it change?

- Can you make up a different story using the same pattern of number sentences?

- What kind of pattern do you get when you change the first addend? The second addend? How are the patterns alike?

▦ How can the pattern you recorded be represented on the calculator?

▦ What do the numbers you are pressing on your calculator represent in your story?

▦ Did you stop using the calculator? When?

Patrones activos
de la suma(continuación)

Introducción (continuación)

5. Registre todas las modificaciones de las operaciones en las hojas de registro y discuta los patrones que se desarrollan.

    Ejemplo:
    2 + 1 = 3
    2 + 2 = 4
    2 + 3 = 5
    2 + 4 = 6
    2 + 5 = 7

6. Pida a los alumnos que elijan una situación similar pero que utilicen números más grandes. Luego, con la ayuda de la calculadora, pídales que encuentren los patrones de las operaciones y los escriban en la hoja de registro.

Cómo reunir y organizar los datos

Mientras los alumnos buscan los patrones, formule preguntas tales como:

- ¿Qué sucedía con la suma cada vez que modificaban un sumando? ¿Disminuía o aumentaba? ¿Por qué?

- ¿Me lo podrían representar con los contadores móviles? ¿Sería la historia igual? ¿De qué manera se podría modificar?

- ¿Pueden inventar una historia diferente utilizando el mismo patrón de operaciones?

- ¿Qué tipo de patrón se obtiene cuando modifican el primer sumando? ¿El segundo? ¿En que se parecen los patrones?

⌨ ¿Cómo pueden representar en la calculadora el patrón que registraron?

⌨ ¿Qué parte de la historia representan los números que están oprimiendo en la calculadora?

⌨ ¿Dejaron de usar la calculadora? ¿En qué momento?

Action-Packed Addition
Patterns (continued)

Analyzing Data and Drawing Conclusions

After students have explored several patterns, have them work as a whole group to analyze their results. Ask questions such as:

- How could you describe the pattern you recorded?

- How did your story have to change as your sentences changed?

- What do you think the patterns might be for subtraction stories? Why?

🖩 How did you use the calculator to help you show the action in your story?

🖩 Does the order in which you entered the numbers in your calculator matter to your story? Why or why not?

Continuing the Investigation

Have students select other stories and number sentences and then repeat the sequence.

Patrones activos
de la suma (continuación)

Cómo analizar los datos y sacar conclusiones

Una vez que los alumnos hayan explorado diversos patrones, trate de que en conjunto analicen las hojas de registro y para ayudarlos, formule preguntas tales como:

- ¿Cómo podrían describir el patrón que registraron?

- ¿Cómo se tuvieron que modificar sus historias a medida que cambiaban las operaciones?

- ¿Cómo piensan que serían los patrones para las historias con restas? ¿Por qué?

▦ ¿Cómo utilizaron la calculadora para que los ayudara a representar la acción de su historia?

▦ ¿Resulta importante para la historia el orden en el que se entraron los números en la calculadora? ¿Por qué?

Cómo continuar la investigación

Pida a sus alumnos que elijan otras operaciones y repitan la secuencia.

Name:

Action-Packed Addition Patterns
Recording Sheet

Collecting and Organizing Data

The number sentence with which I started is:

____ **+** ____ **=**

I have circled the number I want to change.

My number sentence pattern is:

____ **+** ____ **=**

____ **+** ____ **=**

____ **+** ____ **=**

____ **+** ____ **=**

____ **+** ____ **=**

____ **+** ____ **=**

____ **+** ____ **=**

The pattern I see is:

My new number sentence is:

____ **+** ____ **=**

I have circled the number I want to change.

My new number sentence pattern is:

____ **+** ____ **=**

____ **+** ____ **=**

____ **+** ____ **=**

____ **+** ____ **=**

____ **+** ____ **=**

____ **+** ____ **=**

____ **+** ____ **=**

The pattern I see is:

Nombre:

Patrones activos de la suma
Hoja de registro

Cómo reunir y organizar los datos

La operación por la que empecé es la siguiente:

Mi nueva operación es:

___ **+** ___ **=**

___ **+** ___ **=**

Marqué con un círculo el número que deseo cambiar.

Marqué con un círculo el número que deseo cambiar.

El patrón de mi operación es:

El patrón de mi operación es:

___ **+** ___ **=**

___ **+** ___ **=**

___ **+** ___ **=**

___ **+** ___ **=**

___ **+** ___ **=**

___ **+** ___ **=**

___ **+** ___ **=**

___ **+** ___ **=**

___ **+** ___ **=**

___ **+** ___ **=**

___ **+** ___ **=**

___ **+** ___ **=**

___ **+** ___ **=**

___ **+** ___ **=**

El patrón que observo es:

El patrón que observo es:

Action-Packed Subtraction Patterns

Math Concepts

- whole numbers
- subtraction
- patterns

Materials

- TI-108, Math Mate™, Math Explorer™
- manipulatives or counters
- **Action-Packed Subtraction Patterns** recording sheets
- pencils

Overview

Students will use manipulatives and calculators to explore what happens when they change one number at a time in a subtraction number sentence. They will also record and describe the pattern that develops.

Introduction

> The **Action-Packed Stories** activity on page 2 should be completed before beginning this activity.

1. Have each student choose one of his or her **Action-Packed Stories** and write the number sentence that goes with it on the recording sheet: 6 - 1 = 5, for example.

   Or, have students read *There Were Ten in a Bed* by Pam Adams or *Five Little Ducks* by Raffi and write number sentences for one of the stories. Then ask them to choose one of those number sentences and write it on their recording sheets. Next, ask them to model the story they chose with manipulatives.

2. Ask students to choose one number in the story to change, and circle the number in the number sentence on their recording sheets.

   Examples:
   In the number sentence 6 - 1 = 5, change 1 to 2; the new number sentence is 6 - 2 = 4.

   In *There Were Ten in a Bed*, 10 - 1 = 9, 9 - 1 = 8, etc.
   In *Five Little Ducks*, 5 - 1 = 4, 4 - 1 = 3, etc.

3. Now, have students model the new number sentence with manipulatives and discuss how the story should change to go with the new number sentence.

4. Have students continue to change the same number, tell the new story it represents, and model it with manipulatives.

Patrones activos de la resta

Conceptos matemáticos

- números enteros
- resta
- patrones

Materiales

- TI-108, Math Mate™, Math Explorer™
- contadores móviles o contadores
- hojas de registro de **Patrones activos de la resta**
- lápices

Resumen

Los alumnos utilizarán contadores móviles y calculadoras para explorar lo que sucede cuando modifican de a uno por vez los números de una resta. También registrarán y describirán el patrón que se desarrolla.

Introducción

> Antes de iniciar esta actividad se deberá haber completado **Historias activas** de la página 2.

1. Pida a los alumnos que elijan una de sus **Historias activas** y que escriban en la hoja de registro la operación que le corresponde: por ejemplo, 6 - 1 = 5.

   También pueden leer *There Were Ten in a Bed* de Pam Adams o *Five Little Ducks* de Raffi y escribir las operaciones de una de las historias. Luego, pídales que elijan una de las operaciones de la historia y que la escriban en la hoja de registro. A continuación, los niños deberán representar la historia con sus contadores móviles.

2. Pida a los alumnos que elijan el número de la historia que desean modificar. Indíqueles que tracen un círculo alrededor del número de la operación en la hoja de registro.

   Ejemplo:
   En la operación 6 - 1 = 5, cambie el 1 por el 2; la nueva operación será 6 - 2 = 4.

   En *There Were Ten in a Bed*, 10 - 1 = 9, 9 - 1 = 8, etc.
   En *Five Little Ducks,* 5 - 1 = 4, 4 - 1 = 3, etc.

3. A continuación, los alumnos deberán representar la nueva operación con los contadores móviles y discutir la manera en que deben modificar la historia para que se adapte a la nueva operación.

4. Pida a sus alumnos que continúen modificando el mismo número, que cuenten la nueva historia que representa y que la modelen con los contadores móviles.

Action-Packed Subtraction
Patterns (continued)

Introduction (continued)

5. Record all of the changes in the number sentences on their recording sheets and discuss the patterns that develop.

    Example:
    6 - 1 = 5
    6 - 2 = 4
    6 - 3 = 3
    6 - 4 = 2
    6 - 5 = 1

6. Have students choose a similar situation using larger numbers. Then have them use their calculators to find number sentence patterns and write them on their recording sheets.

Collecting and Organizing Data

While students are exploring their patterns, ask questions such as:

*   What happened to your third number when you changed your second number each time? Did it get smaller or larger? Why?

*   What would happen if you changed the first number, but kept the second number the same each time? Could you show me with your manipulatives? Would your story stay the same? How might it change?

*   Can you make up a different story using the same pattern of number sentences?

🖩 How can the pattern you recorded be represented on the calculator?

🖩 What do the numbers you are pressing on your calculator represent in your story?

🖩 Did you stop using the calculator? When?

Patrones activos de la resta (continuación)

Introducción (continuación)

5. Anote todas las modificaciones en las operaciones de las hojas de registro y discuta los patrones que se desarrollan.

   Ejemplo:
   6 - 1 = 5
   6 - 2 = 4
   6 - 3 = 3
   6 - 4 = 2
   6 - 5 = 1

6. Pida a los alumnos que elijan una situación similar pero que utilicen números más grandes. Luego, con la ayuda de la calculadora, deberán encontrar los patrones de las operaciones y escribirlos en la hoja de registro.

Cómo reunir y organizar los datos

Mientras los alumnos exploran sus patrones, formule preguntas tales como las siguientes:

- ¿Qué sucedió con el tercer número cada vez que modificaban el segundo número? ¿Aumentaba o disminuía? ¿Por qué?

- ¿Qué sucedería si modificaran el primer número y no modificaran el segundo cada vez? ¿Me lo podrían representar con los contadores móviles? ¿Se modificaría la historia? ¿Cómo podría cambiar?

- ¿Podrían inventar una historia diferente utilizando el mismo patrón de operaciones?

🖩 ¿De qué manera se puede representar en la calculadora el patrón que registraron?

🖩 ¿Qué parte de la historia representan los números que están oprimiendo en la calculadora?

🖩 ¿Dejaron de usar la calculadora? ¿En qué momento?

Action-Packed Subtraction
Patterns (continued)

Analyzing Data and Drawing Conclusions

After students have explored several patterns, have them work as a whole group to analyze their results. Ask questions such as:

- How could you describe the pattern you recorded?

- How did your story have to change as your sentences changed?

- What do you think the patterns might be for addition stories? Why?

▦ How did you use the calculator to help you show the action in your story?

▦ Does the order in which you entered the numbers in your calculator matter to your story? Why or why not?

Continuing the Investigation

Have students:

- Select other stories and number sentences and repeat the sequence.

- Continue their patterns past $n - n = 0$ and make conjectures about the negative numbers that begin to appear.

Patrones activos de la resta (continuación)

Cómo analizar los datos y sacar conclusiones

Una vez que los alumnos hayan explorado diversos patrones, trate de que en conjunto analicen las hojas de registro y para ayudarlos, formule preguntas tales como las siguientes:

- ¿Cómo podrían describir el patrón que registraron?

- ¿Cómo se tuvo que modificar su historia a medida que cambiaban sus operaciones?

- ¿Cómo piensan que serían los patrones para las historias con sumas? ¿Por qué?

▢ ¿Cómo utilizaron la calculadora para que los ayudara a representar la acción de su historia?

▢ ¿Resulta importante para la historia el orden en el que entraron los números en la calculadora? ¿Por qué?

Cómo continuar la investigación

Pida a sus alumnos que:

- Elijan otras historias y operaciones y repitan la secuencia.

- Continúen con el patrón hasta más allá de $n - n = 0$ y que realicen conjeturas acerca de los números negativos que comienzan a aparecer.

Name:

Action-Packed
Subtraction Patterns
Recording Sheet

Collecting and Organizing Data

The number sentence with which I started is:

____ • ____ =

I have circled the number I want to change.

My number sentence pattern is:

____ • ____ =

____ • ____ =

____ • ____ =

____ • ____ =

____ • ____ =

____ • ____ =

____ • ____ =

____ • ____ =

The pattern I see is:

My new number sentence is:

____ • ____ =

I have circled the number I want to change.

My new number sentence pattern is:

____ • ____ =

____ • ____ =

____ • ____ =

____ • ____ =

____ • ____ =

____ • ____ =

____ • ____ =

____ • ____ =

The pattern I see is:

Nombre:

Patrones activos de la resta

Hoja de registro

Cómo reunir y organizar los datos

La operación por la que empecé es la siguiente:

_____ • _____ = _____

Marqué con un círculo el número que deseo modificar.

El patrón de la operación es:

_____ • _____ =

_____ • _____ =

_____ • _____ =

_____ • _____ =

_____ • _____ =

_____ • _____ =

_____ • _____ =

_____ • _____ =

El patrón que observo es:

Mi nueva operación es:

_____ • _____ = _____

Marqué con un círculo el número que deseo modificar.

Mi nuevo patrón de la operación es:

_____ • _____ =

_____ • _____ =

_____ • _____ =

_____ • _____ =

_____ • _____ =

_____ • _____ =

_____ • _____ =

_____ • _____ =

El patrón que observo es:

Predictable Patterns
with Addition

Math Concepts

- whole numbers
- addition
- comparing numbers

Materials

- TI-108, Math Mate™, Math Explorer™
- **Predictable Patterns with Addition**
  recording sheets
- pencils, crayons, and markers

Overview

Students will generate patterns using repeated addends and different starting points. Then they will analyze and compare the patterns.

Introduction

1. Read *What Comes In 2's, 3's, & 4's?* by Suzanne Aker or *Each Orange Had 8 Slices* by Paul Giganti, Jr.

2. Ask questions such as: If each person has two legs, how many legs are there in the classroom?

3. Discuss with students how they would find the answer to that question (by counting, adding 2 over and over, etc.).

4. Demonstrate how the calculator can help you keep track of adding 2 over and over again by entering **0** ⊞ **2** ⊟⊟⊟⊟ ⊟⊟. . . .

   On a transparency of the hundred chart on the recording sheet, color in the number that is displayed after each press of ⊟. Have students discuss the pattern that is formed.

5. Ask students what kind of pattern they think would be made with **0** ⊞ **3** ⊟⊟⊟⊟⊟⊟. . . ? With **0** ⊞ **4** ⊟⊟⊟ ⊟⊟⊟. . . ? With **0** ⊞ **5** ⊟⊟⊟⊟⊟⊟. . . ? With **1** ⊞ **2** ⊟⊟⊟⊟⊟⊟. . . ? With **1** ⊞ **5** ⊟⊟⊟⊟⊟⊟. . . ?

6. Have students work in pairs or groups. Have them use calculators to generate sequences with repeated addends, recording the numbers in the sequence by coloring them on the hundred chart.

7. Ask students to analyze the patterns they make with different repeated addends and different starting points.

Patrones predecibles de la suma

Conceptos matemáticos

- números enteros
- suma
- comparación numérica

Materiales

- TI-108, Math Mate™, Math Explorer™
- hojas de registro **de Patrones predecibles de la suma**
- lápices, lápices de cera y marcadores de fibra

Resumen

Los alumnos generarán patrones utilizando sumandos repetidos y diferentes puntos de partida. Luego analizarán y compararán los patrones.

Introducción

1. Lea *What Comes In 2's, 3's, and 4's?* de Suzanne Aker o *Each Orange Had 8 Slices* de Paul Giganti, Jr.

2. Formule preguntas tales como: si cada persona tiene dos piernas, ¿cuántas piernas hay en la clase?

3. Discutan la forma en que los alumnos pueden responder a esa pregunta (contando, sumando 2 cada vez, etc.).

4. Demuéstreles la manera en la que la calculadora puede ayudarlos a registrar la suma reiterada de 2 si entran
   0 ⊞ **2** ⊟⊟⊟⊟⊟ . . . .

   Utilice una transparencia de la tabla de la centena en la hoja de registro y coloree el número que aparece en el visor cada vez que se presiona la tecla ⊟. Pida a los niños que debatan acerca del patrón que se forma.

5. Pregunte a los alumnos, ¿qué tipo de patrón resultará con **0** ⊞ **3** ⊟⊟⊟⊟⊟ . . . ? ¿Con **0** ⊞ **4** ⊟⊟⊟⊟⊟ . . . ? ¿Con **0** ⊞ **5** ⊟⊟⊟⊟⊟⊟ . . . ? ¿Con **1** ⊞ **2** ⊟⊟⊟⊟ ⊟⊟ . . . ? ¿Con **1** ⊞ **5** ⊟⊟⊟⊟⊟⊟ . . . ?

6. Pida a los alumnos que trabajen en parejas o grupos. Solicíteles que utilicen sus calculadoras para generar secuencias con sumandos repetidos. Luego, deberán registrar los números de las secuencias coloreándolos en la tabla de la centena.

7. Pida a los alumnos que analicen los patrones que descubrieron cuando repitieron los sumandos y emplearon distintos puntos de partida.

Predictable Patterns
with Addition (continued)

Collecting and Organizing Data

While students are generating data for the different patterns, ask questions such as:

- What addend are you repeating?

- What kinds of objects might you count with that repeated addend?

- What starting point are you using? Why?

- Choose two patterns you made. How are they alike? How are they different?

- Choose a pattern you made that you like. What different repeated addend or starting point could you use to make a pattern similar to it? Try your prediction and see what happens.

- Can you make a pattern with all even numbers? All odd numbers?

🖩 What do you do on the calculator to change the starting point for your pattern?

🖩 What do you do on the calculator to change the repeated addend?

🖩 What happens each time you press ⊟?

Analyzing Data and Drawing Conclusions

After students have made and compared several patterns, have them work as a whole group to analyze their patterns. Ask questions such as:

- Pick two patterns that are alike. How are they alike? Why do you think they turned out alike?

- Pick two patterns that are very different. How are they different? Why do you think they turned out so different?

- What happened when you used the same repeated addend, but started at 1 instead of 0? In what kind of situation might you want to start with 1 instead of 0?

- What happened when you started with other numbers? How did it change your pattern?

🖩 How did you use the calculator in making your patterns?

🖩 Could you predict what the next number on the calculator should be each time?

🖩 How were your patterns related to what you were doing on the calculator?

© 1995 Texas Instruments Incorporated. ™ Trademark of Texas Instruments Incorporated.

Patrones predecibles
de la suma (continuación)

Cómo reunir y organizar los datos (continuación)

Mientras los alumnos generan los datos para los diferentes patrones, formule preguntas tales como:

- ¿Cuál es el sumando que están repitiendo?

- ¿Qué tipo de objetos podrían contar con el sumando que se repite?

- ¿Cuál es el punto de partida que están utilizando? ¿Por qué?

- Elijan dos de los patrones que crearon. ¿En qué se parecen y en qué se diferencian?

- Elijan cualquiera de los patrones que crearon. ¿Qué sumando repetido o punto de partida diferente podrían utilizar para originar un patrón similar al elegido? Inténtenlo y vean lo que sucede.

- ¿Pueden hacer un patrón cuyos números sean todos pares? ¿Y todos números impares?

Cómo analizar los datos y sacar conclusiones

Después de que los alumnos hayan ideado y comparado diversos patrones, trate de que en conjunto los analicen y, para ayudarlos, formule preguntas tales como:

- Elijan dos patrones que se parezcan. ¿En qué se parecen? ¿Por qué piensan que resultaron parecidos?

- Elijan dos patrones que sean muy diferentes entre sí. ¿En qué se diferencian? ¿Por qué piensan que resultaron tan diferentes?

- ¿Qué pasó cuando utilizaron el mismo sumando repetido, pero comenzaron por el 1 en vez de por el 0? ¿En qué situación sería necesario comenzar por el 1 en lugar del 0?

- ¿Qué pasó cuando comenzaron con otros números? ¿De qué manera se modificó el patrón?

- ▦ ¿Qué hacen para modificar el punto de partida de sus patrones en la calculadora?

- ▦ ¿Qué hacen en la calculadora para modificar el sumando repetido?

- ▦ ¿Qué sucede cada vez que oprimen la tecla ⊟ ?

- ▦ ¿Cómo utilizaron la calculadora para generar sus patrones?

- ▦ ¿Pudieron predecir cada vez cuál sería el número que aparecería a continuación en la calculadora?

- ▦ ¿Cómo se relacionaban sus patrones con lo que ustedes hacían en la calculadora?

Predictable Patterns
with Addition (continued)

Analyzing Data and Drawing Conclusions (continued)

- What is alike about the patterns for adding 2 starting with 0 and adding 5 starting with 0? How are they different from other patterns starting with 0? Why do you think they turn out that way?

- How do you make a pattern with only even numbers? Only odd numbers?

Continuing the Investigation

Have students:

- Generate patterns with repeated subtraction starting from 100 and then compare their subtraction patterns to their addition patterns.

- Generate subtraction patterns starting from numbers other than 100 and compare their results to patterns starting from 100.

- Investigate what happens with subtraction patterns when they go "beyond 0" into the negative numbers.

    Note: If students use repeated multiplication, it is important that they know that the TI-108 and Math Mate™ repeat the first factor entered. The Math Explorer™ repeats the second factor.

Patrones predecibles
de la suma (continuación)

Cómo analizar los datos y sacar conclusiones (continuación)

- ¿En qué se parecen los patrones para sumar 2 comenzado por el 0 y para sumar 5 comenzando con el 0? ¿En qué se diferencian de otros patrones que comienzan por el 0? ¿Por qué piensan que es así?

- ¿Cómo se puede hacer un patrón con sólo números pares? ¿Y con sólo números impares?

Cómo continuar la investigación

Indique a sus alumnos que:

- Generen patrones con restas repetidas, comenzando por 100 y que luego comparen sus patrones de restas con los patrones de sumas.

- Generen patrones de restas comenzando por números distintos de 100 y que comparen sus resultados con los patrones que se generaron comenzando por 100.

- Investiguen lo que sucede con los patrones de restas cuando se va "más allá de 0" y comienzan los números negativos.

Nota: Si los alumnos utilizaron la multiplicación repetida, es importante que sepan que la calculadora TI-108 y Math Mate™ repiten el primer factor ingresado. La calculadora Math Explorer™ repite el segundo factor.

Name:

Predictable Patterns with Addition
Recording Sheet

Collecting and Organizing Data

Starting Number   Number (Addend)

_____   _____

Hundred Chart

1	2	3	4	5	6	7	8	9	10
11	12	13	14	15	16	17	18	19	20
21	22	23	24	25	26	27	28	29	30
31	32	33	34	35	36	37	38	39	40
41	42	43	44	45	46	47	48	49	50
51	52	53	54	55	56	57	58	59	60
61	62	63	64	65	66	67	68	69	70
71	72	73	74	75	76	77	78	79	80
81	82	83	84	85	86	87	88	89	90
91	92	93	94	95	96	97	98	99	100

Questions we thought of while we were doing this activity:

Nombre:

Patrones predecibles de la suma
Hoja de registro

Cómo reunir y organizar los datos

Número inicial   **Número (sumando)**

_____   _____

Tabla de la centena

1	2	3	4	5	6	7	8	9	10
11	12	13	14	15	16	17	18	19	20
21	22	23	24	25	26	27	28	29	30
31	32	33	34	35	36	37	38	39	40
41	42	43	44	45	46	47	48	49	50
51	52	53	54	55	56	57	58	59	60
61	62	63	64	65	66	67	68	69	70
71	72	73	74	75	76	77	78	79	80
81	82	83	84	85	86	87	88	89	90
91	92	93	94	95	96	97	98	99	100

Preguntas que surgieron mientras realizábamos esta actividad:

Level 1

Activities:

1   Design a Quilt

2   Double Your Design

3   Symmetry with Pattern Blocks

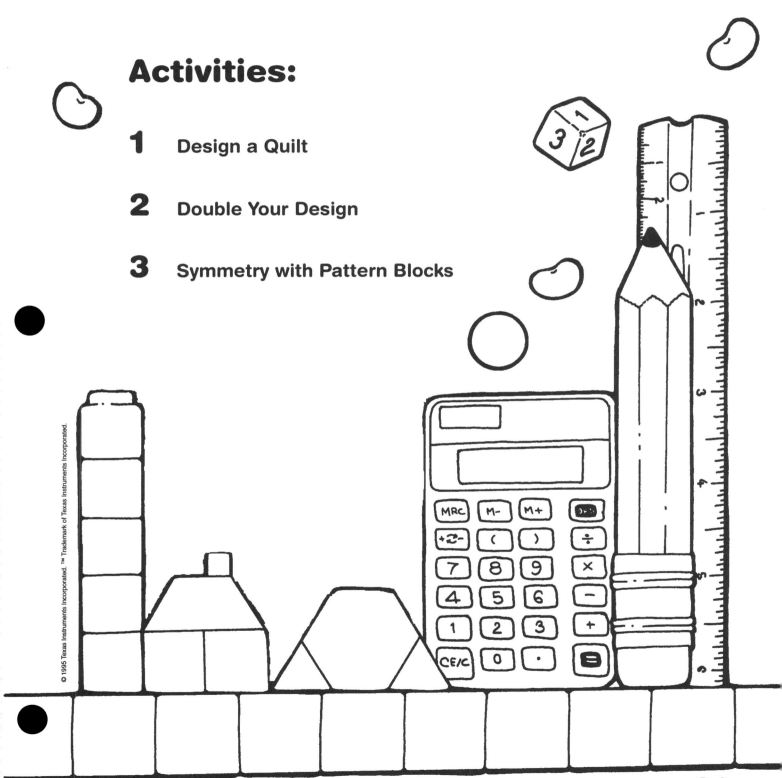

© 1995 Texas Instruments Incorporated. ™ Trademark of Texas Instruments Incorporated.

Design a Quilt

Math Concepts

- whole numbers
- addition
- 2-dimensional geometric figures

Materials

- TI-108, Math Mate™, Math Explorer™
- **Design a Quilt** recording sheets
- Pattern Blocks
- crayons or markers

Overview

Students will use Pattern Blocks to build a design. Then they will use the calculator to determine the value of the design.

Introduction

1. Read a quilting story, such as *The Patchwork Quilt* by Valerie Flournoy, to students and discuss the different geometric figures used in the quilt in the story.

2. Have students build a small quilt design using Pattern Blocks or the paper pattern blocks provided on page 35. Then have them transfer the design to the triangular grid on their recording sheets using crayons or markers.

3. Next, have students assign a value of 1¢ to the green triangle and use the calculator to figure the value of the quilt design.

4. Have each student record the value of the design on the back of their recording sheets, trade designs with another student, and figure the value of their partner's design.

   Note: Other quilting stories include: *The Keeping Quilt* by Patricia Polacco, *Sweet Clara and the Freedom Quilt* by Deborah Hopkinson, and *Sam Johnson and the Blue Ribbon Quilt* by Lisa Ernst.

Collecting and Organizing Data

While students are constructing and recording their quilt designs, ask questions such as:

- What Pattern Blocks are you using in your design? Did any of the blocks seem to not work when you recorded your design on the triangular grid paper? Which ones?

How are you using the calculator to help you find the value of your design?

How can you use Cons on the Math Explorer to help you find the value of your design?

Diseño de una colcha

Conceptos matemáticos

- números enteros
- suma
- figuras geométricas de dos dimensiones

Materiales

- TI-108, Math Mate™, Math Explorer™
- Hojas de registro de **diseño de una colcha**
- Figuras patrón
- Lápices de cera o marcadores de fibra

Resumen

Los alumnos utilizarán figuras patrón para construir un diseño. Luego, utilizarán la calculadora para determinar el valor del diseño.

Introducción

1. Lea a los alumnos una historia de colchas, tal como *The Patchwork Quilt* de Valerie Flournoy, y debatan acerca de las diferentes figuras geométricas utilizadas en la colcha de la historia.

2. Pida a los alumnos que construyan un pequeño diseño de colcha utilizando figuras patrón comerciales o las figuras patrón de papel de la página 35. Pídales que transfieran el diseño a la trama triangular de la hoja de registro utilizando lápices de cera o marcadores de fibra.

3. A continuación, pida a los alumnos que asignen un valor de 1 centavo al triángulo de color verde y que usen la calculadora para calcular el valor del diseño de la colcha.

4. Pida a los alumnos que registren el valor del diseño en la parte posterior de la hoja de registro, cambien diseños con otro alumno y calculen el valor de los diseños de su compañero.

   Nota: Otras historias de colchas son: *The Keeping Quilt* de Patricia Polacco, *Sweet Clara and the Freedom Quilt* de Deborah Hopkinson y *Sam Johnson and the Blue Ribbon Quilt* de Lisa Ernst.

Cómo reunir y organizar los datos

Mientras los niños construyen y registran sus diseños de colcha, formule preguntas tales como:

- ¿Qué figuras patrón están utilizando en su diseño? ¿Parecieron no funcionar algunas figuras cuando registraron su diseño en el papel de trama triangular? ¿Cuáles?

 ¿Cómo están usando la calculadora para encontrar el valor del diseño?

 ¿Cómo pueden usar Cons del Math Explorer para encontrar el valor del diseño?

Design a Quilt (continued)

Collecting and Organizing Data (continued)

- How many green triangles do you think it will take to build the red trapezoid shape? Try it and see. How about the blue parallelogram? The yellow hexagon? Predict first and then find out.

- If we assign a value of 1¢ to the green triangle, how much do you think your whole design is worth? How could you find out? Is there any other way to find the value of your design? Think of one other way, try it, and see if your value remains the same.

▤ What operations can you use on the calculator to help you find the value of your design?

▤ How can you decide if the answer you are getting on the calculator is reasonable or not?

Analyzing Data and Drawing Conclusions

After students have recorded the value of their designs, have them work as a whole group to analyze their triangular grids. Ask questions such as:

- How did you predict the number of green triangles it would take to build the other blocks? Your whole design?

- How could you describe the way you found the value of your design?

- Whose design has the greatest value? The smallest value?

- How many methods of finding the value of your design did you try? Did your value remain the same each time? If not, why do you think it changed?

▤ How did you use the calculator to help you find the value of your design?

▤ What operations did you use on the calculator to help you find the value of your design? Which one do you think worked the best?

▤ Does the order in which you entered the numbers in your calculator matter? Why or why not?

▤ If you changed the value of the green triangle to 2¢, how would you change the way you used the calculator? What would stay the same?

Continuing the Investigation

Have students:

- Change the value of the green triangle and find the new value of their designs.

  Example: If your design was worth 45¢ when the green triangle had a value of 1¢, how much do you think it would be worth if the green triangle had a value of 5¢? Predict first and then find out. If you made a quilt that used this same design nine times, how much would the whole quilt be worth?

© 1995 Texas Instruments Incorporated. ™ Trademark of Texas Instruments Incorporated.

Diseño de una colcha (continuación)

Cómo reunir y organizar los datos (continuación)

- ¿Cuántos triángulos verdes creen que necesitarán para construir la forma trapezoidal de color rojo? Inténtenlo y observen. ¿Qué sucede con el pararalelogramo de color azul? ¿Qué sucede con el hexágono amarillo? Predigan primero y luego descúbranlo.

- Si se asigna un valor de 1 centavo al triángulo verde, ¿cuánto creen que vale la totalidad del diseño? ¿Cómo se puede buscar este valor? ¿Existe otra manera de conocer el valor del diseño? Piensen algún otro método posible, pruébenlo y confirmen si el valor permanece igual.

> ¿Qué operaciones se pueden usar en la calculadora para buscar el valor del diseño?

> ¿Cómo puedes determinar si la respuesta de la calculadora es razonable?

Cómo analizar los datos y sacar conclusiones

Cuando los alumnos hayan registrado el valor de sus diseños, pídales que analicen sus patrones triangulares en conjunto utilizando preguntas tales como:

- ¿Cómo pudieron predecir el número de triángulos verdes que serían necesarios para construir las otras figuras? ¿El diseño completo?

- ¿Cómo podrían describir el modo en que encontraron el valor del diseño?

- ¿Quién creó el diseño con el mayor valor? ¿Con el menor valor?

- ¿Cuántos métodos probaron para encontrar el valor de su diseño? ¿Permaneció sin modificaciones el valor? En caso contrario, ¿por qué creen que cambió?

> ¿Cómo utilizaron la calculadora para encontrar el valor del diseño?

> ¿Qué operaciones de la calculadora utilizaron para encontrar el valor del diseño? ¿Qué operación creen funcionó mejor?

> ¿Es importante el orden en el que se entran los números a la calculadora? ¿Por qué?

> ¿Si se cambia el valor del triángulo verde a 2 centavos, ¿cambiaría esto la forma en que se usa la calculadora? ¿Qué permanecería sin modificaciones?

Cómo continuar la investigación

Indique a sus alumnos que:

- Cambien el valor del triángulo verde y encuentren el nuevo valor de sus diseños.
  Ejemplo: Si el diseño valía 45 centavos cuando al triángulo verde le correspondía un valor de 1 centavo, ¿cuánto creen que valdría si el triángulo verde tuviera un valor de 5 centavos? Predigan primero y luego investiguen. Si construyeran una colcha en la que se usa este mismo diseño nueve veces, ¿cuánto valdría la totalidad de la colcha?

Name:

Design a Quilt
Recording Sheet

Collecting and Organizing Data

Record your quilt design below.

Analyzing Data and Drawing Conclusions

Record the value of your design on the back of this paper.

Questions we thought of while we were doing this activity:

Nombre:

Diseño de una colcha
Hoja de registro

Cómo reunir y organizar los datos

A continuación, registre su diseño de colcha.

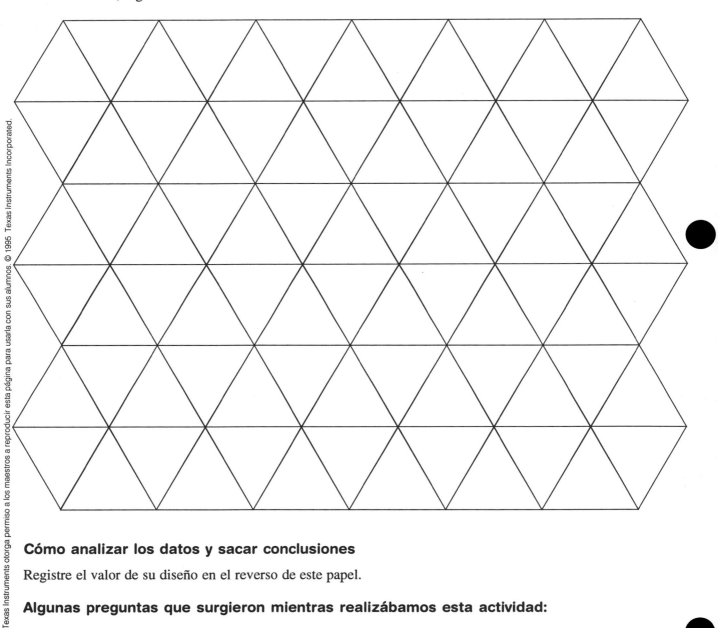

Cómo analizar los datos y sacar conclusiones

Registre el valor de su diseño en el reverso de este papel.

Algunas preguntas que surgieron mientras realizábamos esta actividad:

Design a Quilt

Pattern Blocks

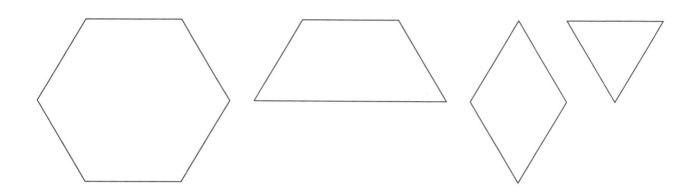

Other Geometric Shapes

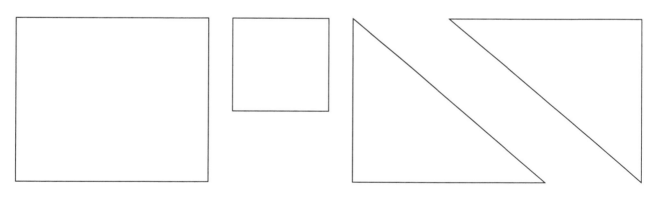

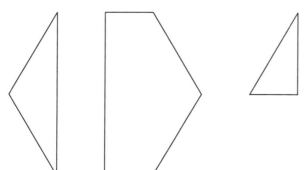

Diseño de una colcha

Figuras patrón

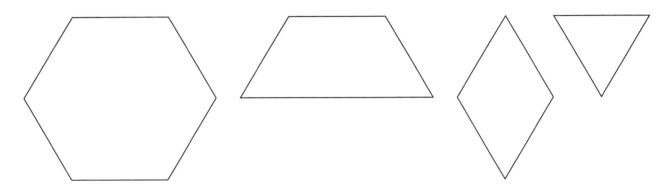

- -

Otras formas geométricas

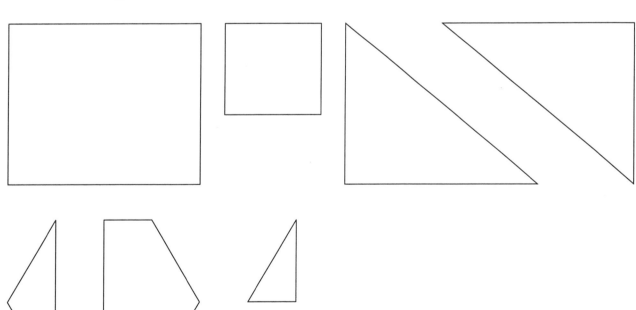

Double Your Design

TI-108
MATH MATE
MATH EXPLORER

Math Concepts

- whole numbers
- addition
- 2-dimensional geometric figures
- symmetry

Materials

- TI-108, Math Mate™, Math Explorer™
- Pattern Blocks
- **Double Your Design** recording sheets
- small mirrors
- crayons or markers

Overview

Students will use Pattern Blocks to investigate symmetry by building a design, making its mirror image, and using the calculator to determine the value of both the first design and its reflection.

Introduction

Completing the **Design a Quilt** activity on page 32 before beginning this activity might be helpful.

1. On the overhead projector, build a simple design with overhead Pattern Blocks. Decide on a line of symmetry, mark it with a transparency marker, and use additional Pattern Blocks to build the mirror image of the original design.

   Note: A small mirror is helpful in deciding where the blocks should be placed to form a reflection of the original design.

2. Have students build their own designs using Pattern Blocks or the paper pattern blocks provided on page 39. Then have them transfer their designs to the triangular grid on their recording sheets, using crayons or markers.

3. Ask students to assign a value of 1¢ to the green triangle and figure the value of their original designs.

4. Have students show the reflections of their designs on the triangular grid on their recording sheets.

5. Have students use the value of their original designs to predict the value of their reflected designs. Ask them to record their predictions.

6. Then have students figure the total value of their reflected designs and compare it to their predictions.

Duplica tu diseño

Conceptos matemáticos

- números enteros
- suma
- figuras geométricas de dos dimensiones
- simetría

Materiales

- TI-108, Math Mate™, Math Explorer™
- figuras patrón
- hojas de registro de **Duplica tu diseño**
- espejos pequeños
- lápices de cera o marcadores de fibra

Resumen

Los alumnos utilizarán las figuras patrón para investigar simetrías construyendo un diseño, realizarán su imagen especular y usarán la calculadora para determinar el valor de su primer diseño y su reflejo.

Introducción

Antes de iniciar esta actividad es conveniente haber completado **Diseño de una colcha** de la página 32.

1. En el proyector de transparencias, construya un diseño simple con las figuras patrón proyectadas. Decídase por una línea de simetría, márquela con un marcador de transparencias y utilice las figuras patrón adicionales para construir la imagen especular del diseño original.

   Nota: Es conveniente utilizar un pequeño espejo para determinar la posición de las figuras para formar el reflejo del diseño original.

2. Pida a los alumnos que construyan sus propios diseños utilizando bloques patrón comerciales o los bloques patrón de papel que se proporcionan en la página 39. Luego, pídales que transfieran sus diseños al papel con trama triangular en la hoja de registro usando lápices de cera o marcadores de fibra.

3. A continuación, pida a los alumnos que asignen un valor de 1 centavo al triángulo de color verde y que calculen el valor de su diseño original.

4. Pida a los alumnos que muestren el reflejo de su diseño en la trama triangular de la hoja de registro.

5. Pida a los alumnos que empleen el valor de sus diseños originales para predecir el valor de sus diseños reflejados. Pídales que registren sus predicciones.

6. Luego, solicite a los alumnos que calculen el valor total de sus diseños reflejados y que lo comparen con sus predicciones.

Double Your Design (continued)

Collecting and Organizing Data

While students are constructing and recording their designs, ask questions such as:

- What Pattern Blocks are you using in your design?

- Where is your line of symmetry? How are you using the mirror to reflect your design along your line of symmetry?

- If we assign a value of 1¢ to the green triangle, how much do you think your original design is worth? Record your prediction and then find out.

- How much do you think your reflected design is worth? Write down your prediction and then find out.

- What is the total value of both designs? Write down your prediction and then find out.

> How can you use [Cons] on the Math Explorer to help you find the value of your design?
>
> How are you using the calculator to help you find the value of your design?
>
> What operations can you use on the calculator to help you find the value of your design?
>
> How can you decide if the answer you are getting on the calculator is reasonable or not?

Analyzing Data and Drawing Conclusions

After students have recorded the value of their designs, have them work as a whole group to analyze their triangular grids. Ask questions such as:

- How did you predict the number of green triangles it would take to build your original design?

- How could you describe the way you found the value of your original design? Its reflection?

- How did the mirror help you build the reflection of your original design?

- How did you find the total value of your design and its reflection?

> How did you use the calculator to help you find the value of your designs?
>
> What operations did you use on the calculator to help you find the value of your designs? Which one do you think worked the best?
>
> Does the order in which you entered the numbers in your calculator matter? Why or why not?
>
> If you changed the value of the green triangle to 2¢, how would you change the way you used the calculator? What would stay the same?

Continuing the Investigation

Have students change the value of the green triangle and find the new values of their designs.

Duplica tu diseño (continuación)

Cómo reunir y organizar los datos

Mientras los alumnos construyen y registran sus diseños, formule preguntas tales como:

- ¿Qué figuras patrón utilizan en sus diseños?

- ¿Dónde está situada su línea de simetría? ¿Cómo utilizan el espejo para reflejar el diseño junto a la línea de simetría?

- Si asignamos el valor de 1 centavo al triángulo de color verde, ¿cuánto creen que valdría el diseño original? Registren sus predicciones y luego calcúlenlo.

- ¿Cuánto creen que vale el diseño reflejado? Registren sus predicciones y luego averíguenlas.

- ¿Cuál es el valor total de ambos diseños? Registren sus predicciones y luego averíguenlas.

- ¿De qué manera se puede usar Cons en el Math Explorer para conocer el valor del diseño?

- ¿De qué modo utilizas la calculadora para conocer el valor del diseño?

- ¿Qué operaciones se pueden utilizar en la calculadora para conocer el valor del diseño?

- ¿Cómo pueden determinar si es razonable la respuesta que aparece en la calculadora?

Cómo analizar los datos y sacar conclusiones

Cuando los alumnos hayan registrado el valor de sus diseños, pídales que analicen sus tramas triangulares en grupo, y formúleles preguntas tales como:

- ¿De qué modo pudieron predecir el número de triángulos verdes que necesitarían para construir su diseño original?

- ¿Cómo describirían el método mediante el cual averiguaron el valor del diseño original? ¿El de su reflejo?

- ¿Cómo facilitó el espejo la construcción del reflejo del diseño original?

- ¿Cómo conocieron el valor total del diseño y de su reflejo?

- ¿Cómo utilizaron la calculadora para encontrar el valor de los diseños?

- ¿Qué operaciones de la calculadora utilizaron para encontrar el valor de los diseños? ¿Qué operación creen funcionó mejor?

- ¿Es importante el orden en el que se ingresan los números a la calculadora? ¿Por qué?

- Si se cambia el valor del triángulo verde a 2 centavos, ¿cambiaría esto la forma en que se usa la calculadora? ¿Qué permanecería sin modificaciones?

Cómo continuar la investigación

Pida a los alumnos que cambien el valor del triángulo verde y que encuentren los nuevos valores de sus diseños.

Name:

Double Your Design
Recording Sheet

Collecting and Organizing Data

Record your design and its mirror image below. Be sure to include your line of symmetry.

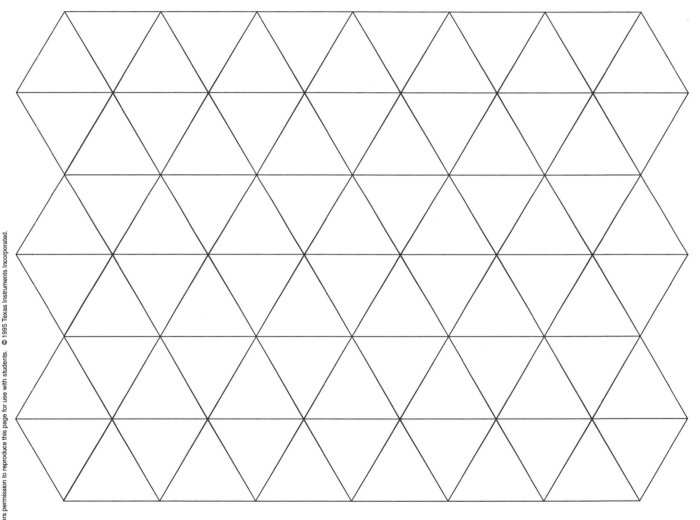

Analyzing Data and Drawing Conclusions

My first design is worth:_____

My design and its reflection is worth:_____

Questions we thought of while we were doing this activity:

Nombre:

Duplica tu diseño
Hoja de registro

Cómo reunir y organizar los datos

A continuación, registre su diseño y su imagen reflejada. Asegúrese de incluir la línea de simetría.

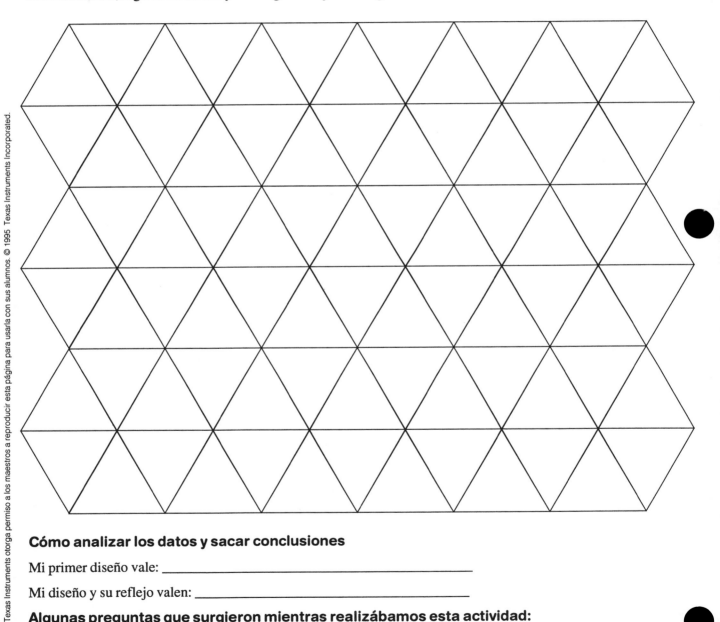

Cómo analizar los datos y sacar conclusiones

Mi primer diseño vale: _____

Mi diseño y su reflejo valen: _____

Algunas preguntas que surgieron mientras realizábamos esta actividad:

Double Your Design

Pattern Blocks

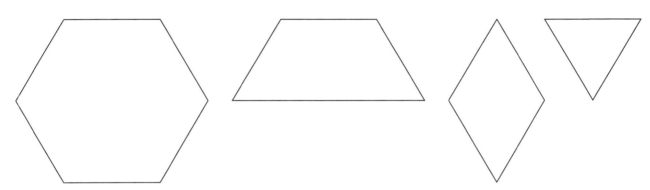

- -

Other Geometric Shapes

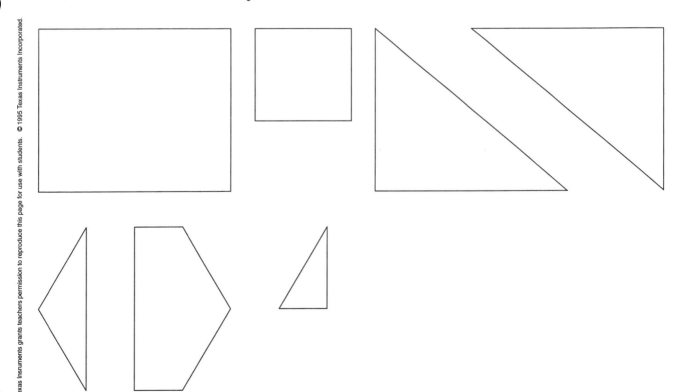

Duplica tu diseño

Figuras patrón

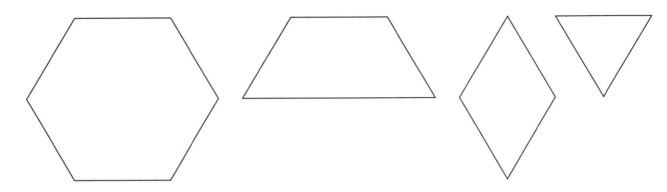

Otras formas geométricas

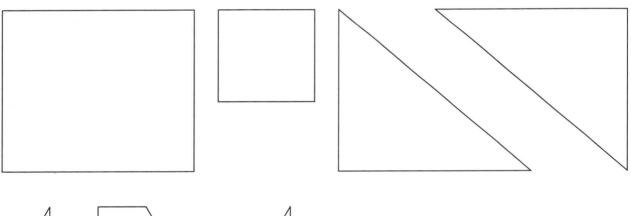

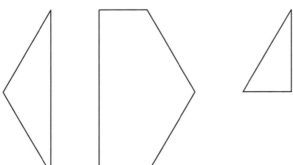

Symmetry with Pattern Blocks

Math Concepts

- whole numbers
- addition
- 2-dimensional geometric figures
- symmetry

Materials

- TI-108, Math Mate™, Math Explorer™
- **Symmetry with Pattern Blocks** recording sheets
- small mirrors
- Pattern Blocks
- crayons or markers

Overview

Students will use Pattern Blocks to build a design that has a line of symmetry and find the value of half of the design. Finally they will predict and then discover the value of the entire design.

Introduction

> The **Design a Quilt** (page 32) and **Double Your Design** (page 36) activities should be completed before beginning this activity.

1. On the overhead projector, build a design that has a line of symmetry using Pattern Blocks. Mark the line of symmetry with a transparency marker, and check it with a small mirror. Ask students to predict the value of half of the design if the green triangle is worth 1¢. Find the value of half of the design. Have students use that information to predict the value of the entire design.

2. Have students build their own designs using Pattern Blocks or the paper pattern blocks provided on page 43. Have them record their designs on the triangular grid paper provided on their recording sheets using crayons or markers. Have them repeat the process modeled on the overhead: Predict the value of half of the design, find the value, predict the value of the whole design, and find its value.

3. Have students write what they discovered from the three activities: **Design a Quilt**, **Double Your Design**, and **Symmetry with Pattern Blocks**.

Collecting and Organizing Data

While students are constructing and recording their designs, ask questions such as:

- What Pattern Blocks are you using in your design?

 How are you using the calculator to help you find the value of your design?

Simetría con las figuras patrón

Conceptos matemáticos

- números enteros
- suma
- figuras geométricas de dos dimensiones
- simetría

Materiales

- TI-108, Math Mate™, Math Explorer™
- hojas de registro de **Simetría con las figuras patrón**
- espejos pequeños
- figuras patrón
- lápices de cera o marcadores de fibra

Resumen

Los alumnos utilizarán las figuras patrón para construir un diseño con una línea de simetría y encontrar el valor de la mitad del diseño. Finalmente, predecirán y encontrarán el valor de la totalidad del diseño.

Introducción

> Antes de iniciar esta actividad se debe haber completado **Diseño de una colcha** (página 32) y **Duplica tu diseño** (página 36).

1. En el proyector de transparencias, construya un diseño con las figuras patrón con una línea de simetría. Marque esta línea con un marcador de transparencias y verifíquela con un pequeño espejo. Pida a los alumnos que predigan el valor de la mitad del diseño si el valor del triángulo de color verde es de 1 centavo. Encuentren el valor de la mitad del diseño. Solicite a los alumnos que empleen esa información para predecir el valor de la totalidad del diseño.

2. Pida a los alumnos que construyan sus propios diseños utilizando las figuras patrón comerciales o las figuras patrón de papel que se proporcionan en la página 43. Solicíteles que registren sus diseños en el papel con trama triangular de la hoja de registro usando lápices de cera o marcadores. Pídales que repitan el proceso modelado en el proyector de transparencias, que predigan el valor de la mitad del diseño y el valor de la totalidad del diseño, y que encuentren su valor.

3. Pida a los alumnos que escriban lo que descubrieron en las tres actividades: **Diseño de una colcha**, **Duplica tu diseño** y **Simetría con las figuras patrón**.

Cómo reunir y organizar los datos

Mientras los niños construyen y registran sus diseños, formule preguntas tales:

- ¿Qué figuras patrón están utilizando en su diseño?

 ¿Cómo están usando la calculadora para encontrar el valor del diseño?

Symmetry with Pattern Blocks (continued)

Collecting and Organizing Data (continued)

- Where is your line of symmetry? How are you using the mirror to check your line of symmetry?

- If we assign a value of 1¢ to the green triangle, how much do you think half of your design is worth? How can you find out?

- How can you use the value of half of your design to predict how much your whole design is worth? Write down your prediction and then find the value.

How can you decide if the answer you are getting on the calculator is reasonable or not?

Analyzing Data and Drawing Conclusions

After students have recorded the value of their designs, have them work as a whole group to analyze their triangular grids. Ask questions such as:

- How did you predict the number of green triangles it would take to build half of your design?

- Did the line of symmetry in your design divide any of your Pattern Blocks into parts? How did you decide to count the value of those parts?

- How did the mirror help you check the line of symmetry in your design?

How did you use the calculator to help you find the value of your design?

Does the order in which you entered the numbers in your calculator matter? Why or why not?

Continuing the Investigation

Have students:

- Change the value of the green triangle and find the new values of their designs.

- Make a design with two lines of symmetry.

Simetría con las figuras patrón (continuación)

Cómo reunir y organizar los datos (continuación)

- ¿Dónde está situada la línea de simetría? ¿Cómo se utiliza el espejo para verificar la línea de simetría?

- Si se asigna el valor de 1 centavo al triángulo verde, ¿cuánto creen que vale la mitad del diseño? ¿Cómo se puede buscar este valor?

- ¿Cómo se puede usar el valor de la mitad del diseño para predecir cuánto vale la totalidad del mismo? Registren primero sus predicciones y luego busquen el valor.

▣ ¿Cómo es posible determinar si la respuesta de la calculadora es razonable?

Cómo analizar los datos y sacar conclusiones

Cuando los alumnos hayan registrado el valor de sus diseños, pídales que analicen sus patrones triangulares en conjunto. Utilice preguntas tales como:

- ¿Cómo pudieron predecir el número de triángulos verdes que serían necesarios para construir la mitad del diseño?

- ¿Dividió la línea de simetría del diseño alguna de las figuras patrón? ¿De qué manera decidieron contar el valor de esas partes?

- ¿De qué modo facilitó el espejo la verificación de la línea de simetría del diseño?

▣ ¿Cómo utilizaron la calculadora para encontrar el valor del diseño?

▣ ¿Es importante el orden en el que se ingresan los números a la calculadora? ¿Por qué?

Cómo continuar la investigación

Indique a sus alumnos que:

- Cambien el valor del triángulo verde y encuentren el nuevo valor de sus diseños.

- Hagan un diseño con dos líneas de simetría.

Name: _____

Symmetry with Pattern Blocks
Recording Sheet

Collecting and Organizing Data

Record your design below. Be sure to include your line of symmetry.

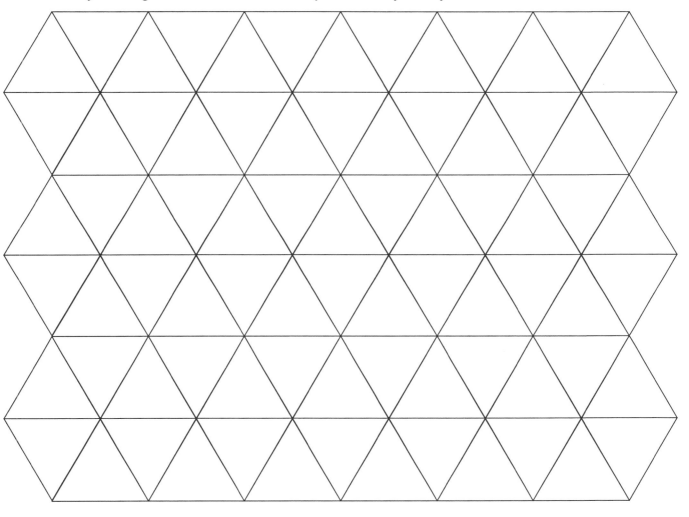

Analyzing Data and Drawing Conclusions

Half of my design is worth: _____ (my prediction) _____ (actual value)

My whole design is worth: _____ (my prediction) _____ (actual value)

Questions we thought of while we were doing this activity:

Nombre:

Simetría con las figuras patrón
Hoja de registro

Cómo reunir y organizar los datos

A continuación, registre su diseño. No olvide incluir la línea de simetría.

Cómo analizar los datos y sacar conclusiones

La mitad de mi diseño vale: _____ (mi predicción) _____ (valor real)

La totalidad de mi diseño vale: _____ (mi predicción) _____ (valor real)

Algunas preguntas que surgieron mientras realizábamos esta actividad:

Symmetry with Pattern Blocks

Pattern Blocks

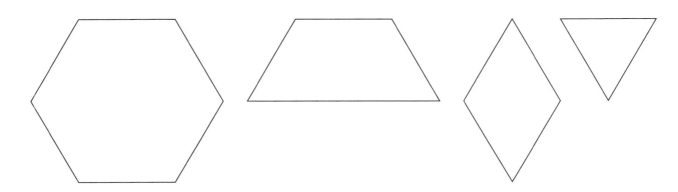

Other Geometric Shapes

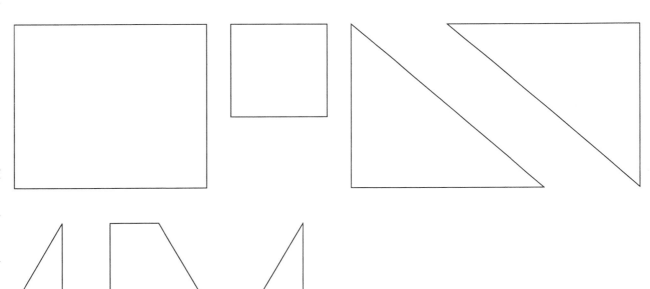

Simetría con las figuras patrón

Figuras patrón

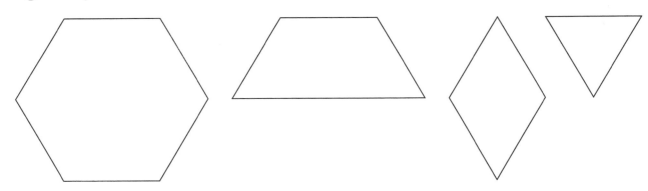

- -

Otras formas geométricas

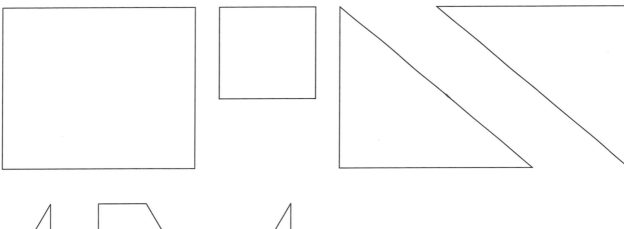

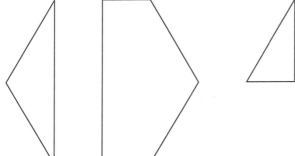

Probability and Statistics

Level 1

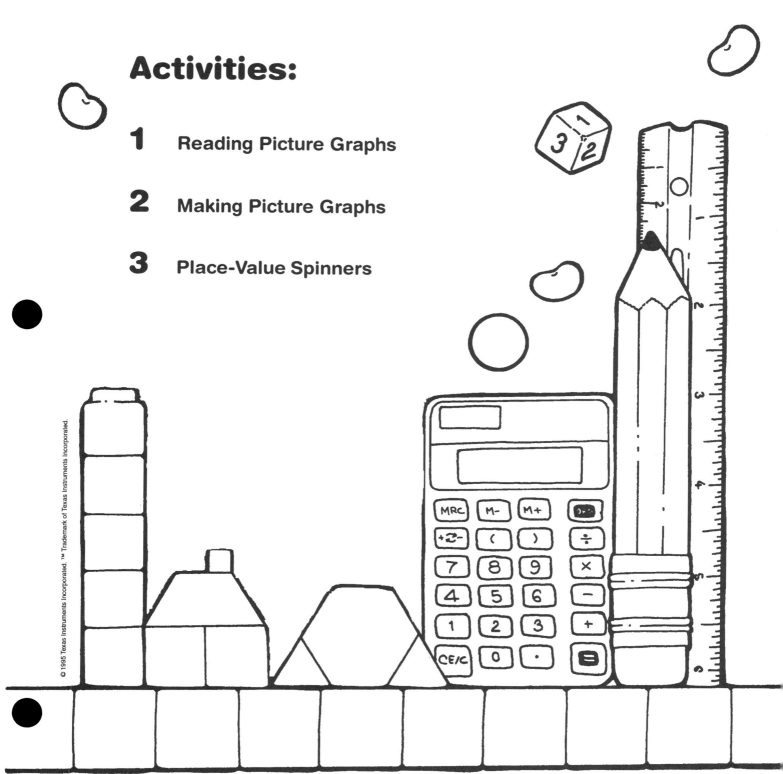

Activities:

1 Reading Picture Graphs

2 Making Picture Graphs

3 Place-Value Spinners

Reading Picture Graphs

Math Concepts
- whole numbers
- patterns
- addition
- multiplication
- graphing

Materials
- Math Explorer™
- **Reading Picture Graphs** recording sheets
- colored chips or tiles
- copies of appropriate picture graphs

Overview

Students will use calculators to help them read picture graphs.

Introduction

1. Talk with students about collecting data for a picture graph.

   Example: Suppose every student in the school voted for their favorite subject. How large do you think the graph would be if every piece of data became an individual picture on the picture graph?

2. Discuss putting the students' votes together in groups of 5, 8, 10, or 20 and the effect that each grouping would have on the graph.

3. Give students a sample picture graph in which the pictures represent more than one piece of data. (The picture graph on page 48 is an example.)

4. Using the sample graph, have students translate the pictures on the graph back into pieces of data by placing the appropriate number of colored chips or tiles directly on each of the pictures.

5. Then have students use the [Cons] key to count the chips and look for patterns.

   Example: If each picture represents 5 pieces of data, students should:

   a. Place 5 chips on each picture.

   b. Separate the chips on the first picture, laying them beside the graph to clearly show the 5 pieces of data that each picture represents.

   c. Enter [+] **5** [Cons] **0** to prepare the calculator to count by 5s.

Cómo leer gráficos con dibujos

Conceptos matemáticos

- números enteros
- patrones
- suma
- multiplicación
- representaciones gráficas

Materiales

- Math Explorer™
- hojas de registro de **Cómo leer gráficos con dibujos**
- fichas de colores
- copias de gráficos con dibujos apropiados

Resumen

Los alumnos utilizarán la calculadora para que ésta les facilite la lectura de los gráficos con dibujos.

Introducción

1. Hable con los alumnos acerca de la reunión de datos para un gráfico con dibujos.

    Ejemplo: Suponga que cada alumno de la escuela vota su tema preferido. ¿Qué dimensiones creen que tendrá el gráfico si cada dato se convierte en un dibujo en el gráfico?

2. Debatan si se agrupan los votos de 5, 8, 10 ó 20 estudiantes y el efecto que cada agrupamiento tendrá en el gráfico.

3. Ofrezca a los alumnos un gráfico de dibujos de muestra, en el cual los dibujos representen más de un dato. (En la página 56 encontrará un ejemplo de gráfico de dibujos.)

4. Utilice el gráfico de ejemplo y pida a los alumnos que traduzcan los dibujos del gráfico nuevamente a información. Para hacerlo han de colocar el número apropiado de fichas de color directamente sobre cada uno de los dibujos.

5. A continuación, pida a los alumnos que utilicen la tecla [Cons] para contar la fichas y buscar patrones.

    Ejemplo: Si cada dibujo representa cinco datos, los alumnos deberán:

    a. Situar 5 fichas en cada dibujo.

    b. Separar las fichas del primer dibujo, distribuyéndolas junto al gráfico para mostrar claramente los 5 datos que representa cada dibujo.

    c. Entrar [+] **5** [Cons] **0** para preparar la calculadora para contar de 5 en 5.

Reading Picture Graphs (continued)

Introduction (continued)

d. Enter [Cons] to count the first group of 5. The counter **1** and the result **5** show in the calculator display.

e. Separate the chips on the second picture, enter [Cons] again to count the second group of 5, see **2** and **10** in the display, and so on.

Collecting and Organizing Data

While students are using the [Cons] function to count the data in a picture graph, ask questions such as:

- What do the pictures in the picture graph represent? How do you know that?

- What do your colored chips represent?

- How can you tell by looking at the graph which categories have more data? Less data? The same amount of data?

- Do your strategies for finding more, less, and about the same amount of data need to change if the pictures represent more than one piece of information? Why or why not?

How can you use the calculator to help you read the data?

How can you use [Cons] on the Math Explorer to help you read the data? How do you decide what to enter before pressing [Cons]? Why do you begin at 0?

How can you use [=] to help you read the data?

Analyzing Data and Drawing Conclusions

After students have read their graphs, have them discuss the graphs as a whole group. Ask questions such as:

- What does the data on the graph tell you?

- What is a question that could be answered by this graph?

- What is a question that could not be answered by this graph?

- Why do you think the designers of this graph chose each of their pictures to stand for ____ number of _____?

- What are the advantages of having a picture represent more than one piece of data?

- What are the disadvantages of having a picture represent more than one piece of data?

How did you use [Cons] to read the graph?

How could you use [=] to read the graph?

How did the colored chips laid beside your graph as you counted them connect to the display you saw on your calculator?

Cómo leer gráficos con dibujos (continuación)

Introducción (continuación)

d. Entrar [Cons] para contar el primer grupo de 5. El contador **1** y el resultado **5** aparecen en el visor de la calculadora.

e. Separar las fichas en el segundo dibujo, entrar [Cons] nuevamente para contar el segundo grupo de 5, ver **2** y **10** en el visor de la calculadora, y así sucesivamente.

Cómo reunir y organizar los datos

Mientras los alumnos utilizan la función [Cons] para contar los datos de un gráfico con dibujos, formule preguntas tales como:

- ¿Qué representan los dibujos del gráfico? ¿Cómo lo saben?

- ¿Qué representan las fichas de color?

- ¿Cómo pueden determinar a partir del gráfico qué categorías tienen más datos? ¿Menos datos? ¿La misma cantidad de datos?

- ¿Es necesario que cambien sus estrategias para encontrar más, menos o la misma cantidad de datos si los dibujos representan más de un dato? ¿Por qué?

⊞ ¿Cómo pueden utilizar la calculadora para leer los datos?

⊞ ¿Cómo pueden utilizar la función [Cons] del Math Explorer para leer los datos? ¿Cómo determinar qué entrar antes de oprimir [Cons]? ¿Por qué comienzan con 0?

⊞ ¿Cómo pueden usar ⊟ para facilitar la lectura de los datos?

Cómo analizar los datos y sacar conclusiones

Una vez que los alumnos hayan leído sus gráficos, pídales que discutan los gráficos en grupo formulándoles preguntas tales como las siguientes:

- ¿Qué indican los datos del gráfico?

- ¿A qué pregunta se puede responder mediante este gráfico?

- ¿A qué pregunta no se puede responder mediante este gráfico?

- ¿Creen que las personas que diseñaron este gráfico eligieron cada uno de los dibujos para representar _____ veces _____?

- ¿Cuáles son las ventajas de que un dibujo represente más de un dato?

- ¿Cuáles son las desventajas de que un dibujo represente más de un dato?

⊞ ¿Cómo usó la función [Cons] para leer el gráfico?

⊞ ¿Cómo puede usar ⊟ para leer el gráfico?

⊞ ¿De qué modo se relacionaron las fichas de color situadas junto al gráfico con la información que aparecía en el visor de la calculadora, a medida que las contaba?

Reading Picture Graphs (continued)

Continuing the Investigation

Have students:

- Read and discuss a picture graph that has half of a picture in one of its categories.

- Collect samples of picture graphs from magazines and the newspaper and evaluate them based on their clarity and appeal. **Note:** *Weekly Reader* and *USA Today* are good sources of picture graphs.

Cómo leer gráficos con dibujos (continuación)

Cómo continuar la investigación

Indique a sus alumnos que:

- Lean y discutan un gráfico de dibujos que tenga en una de sus categorías la mitad de un dibujo.

- Reúna muestras de gráficos con dibujos de revistas y periódicos y evalúen los mismos según su claridad y preferencias. **Nota:** *Weekly Reader* y *USA Today* son convenientes fuentes para gráficos de dibujos.

Name:

Reading Picture Graphs
Recording Sheet

Collecting and Organizing Data

Favorite Fruit

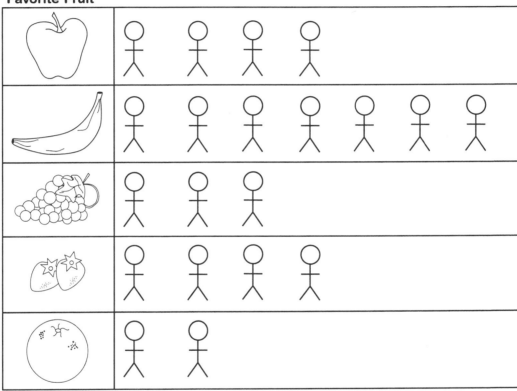

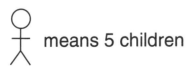 means 5 children

Patterns we found while we were doing this activity:

Nombre:

Cómo leer gráficos con dibujos
Hoja de registro

Cómo reunir y organizar los datos

Fruta favorita

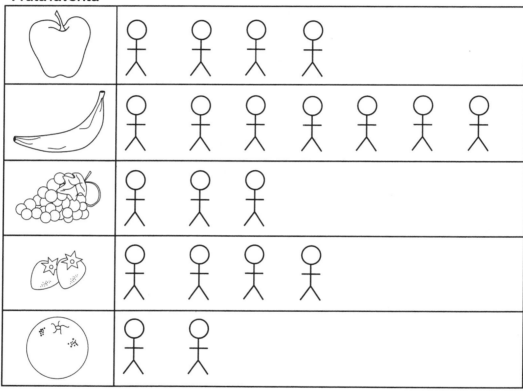

 significa 5 niños

Patrones que encontramos mientras realizábamos esta actividad:

Making Picture Graphs

Math Concepts
- whole numbers
- multiplication
- patterns
- graphing
- addition

Materials
- Math Explorer™
- **Making Picture Graphs** recording sheets
- 3- x 5-inch note cards
- colored chips or linking cubes
- pencils or markers

Overview

Students will use calculators to help make picture graphs with symbols that represent more than one piece of data.

Introduction

> The **Reading Picture Graphs** activity on page 45 should be completed before beginning this activity.

1. Have students work as a whole class to brainstorm a list of things they would like to know about other people. Some books that might stimulate ideas are *These Are My Pets* and *When I Get Bigger*, both by Mercer Mayer.

2. Have students work in groups of four to choose one characteristic about which to collect data. Then have them design a question for collecting the data. Students should:

   a. Include three possible answers to the question from which respondents may choose.

   b. Ask the question orally, or record it on a 3- by 5-inch card for respondents to read.

   c. Present the question and possible answers to other students for their responses.

   d. Record and tally the responses on their recording sheets.

   Note: Encourage students to be creative in designing their question cards.

3. Have each student in the small groups collect at least 20 pieces of data so that each group has a total of 80 pieces of data. Have students represent the data with colored chips or linking cubes.

4. Discuss putting the data in groups of 5, 8, 10, or 20 and the effect that each grouping might have on the look of the graph.

   Note: Students might even experiment by grouping the chips or cubes they used to represent their data.

© 1995 Texas Instruments Incorporated. ™ Trademark of Texas Instruments Incorporated.

Creación de gráficos con dibujos

Conceptos matemáticos

- números enteros
- multiplicación
- patrones
- representaciones gráficas
- suma

Materiales

- Math Explorer™
- hojas de registro de **Creación de gráficos con dibujos**
- tarjetas de cartón de 3 x 5 pulgadas
- fichas de colores o cubos de enlace
- lápices o marcadores de fibra

Resumen

Los alumnos utilizarán la calculadora para que ésta les facilite la creación de gráficos con símbolos que representan más de un dato.

Introducción

Antes de iniciar esta actividad se debe haber completado **Cómo leer gráficos con dibujos** en la página 45.

1. Pida a los alumnos que trabajen todos juntos para crear una lista de las cosas que les gustaría conocer acerca de otras personas. Algunos de los libros que pueden estimular ideas sobre estos temas son: *These Are My Pets* y *When I Get Bigger,* ambos de Mercer Mayer.

2. Solicite a los alumnos que trabajen en grupos de cuatro para elegir una de las características sobre la cual puedan reunir información. Luego, pídales que diseñen una pregunta para obtener los datos. Los estudiantes deben:

    a. Ofrecer tres respuestas posibles a la pregunta para que los encuestados puedan elegir una de ellas.

    b. Realizar las preguntas en forma verbal, o registrarla en una tarjeta de 3 x 5 pulgadas, para que puedan leerlas las personas encuestadas.

    c. Presentar la pregunta y las posibles respuestas a otros alumnos, para obtener sus respuestas.

    d. Registrar y calcular las respuestas en la hoja de registro.

    Nota: Anime a los alumnos a ser creativos en el diseño de sus tarjetas de cuestionario.

3. Pida a cada alumno de los pequeños grupos que reuna por lo menos 20 datos, de modo que el grupo cuente, en total, con 80 datos. Solicite a los alumnos que representen los datos con fichas de color o cubos de enlace.

4. Debata la clasificación de los datos en grupos de 5, 8, 10 ó 20 y el efecto que este agrupamiento podrá tener en el aspecto del gráfico.

    Nota: Los alumnos incluso pueden experimentar agrupando las fichas o los cubos que utilizaron para representar sus datos.

Making Picture Graphs (continued)

Introduction (continued)

5. Then have students use the [Cons] key to experiment with grouping data.

   Example: To explore with groups of 5, have students:

   a. Enter [+] **5** [Cons] **0** to prepare the calculator to count by 5s.
   b. Enter [Cons] to count the first group of 5. The counter **1** and the result **5** are displayed.
   c. Stack five chips or cubes together to represent the first picture in that category on the graph.
   d. Press [Cons] again to count the second group of 5. The counter **2** and the result **10** are displayed.
   e. Stack a second group of five chips or cubes to represent the second picture in that category on the graph.
   f. Continue the process until the data for that category is exhausted (all the chips or cubes are used).

6. Have students record their data on the picture graph template on their recording sheets, using pictures to represent the group size they chose.

Collecting and Organizing Data

While students are using the [Cons] function and their counters to group the data for the picture graph, ask questions such as:

- What do the chips or cubes you are using represent?

- How did you decide to group your data? Why did you decide to group it that way?

- How are you recording these groups on the picture graph?

- What are you going to do when you don't have enough counters left to make another group?

- How are you going to record the leftovers on the picture graph?

- What patterns, if any, are you finding when you group your data?

How can you use the calculator to help you group the data?

How can you use the [Cons] key on the Math Explorer to help you group the data? How do you decide what to enter before pressing [Cons]?

How can you use addition with [Cons] to group your data? Where do you start?

How can you use subtraction with [Cons] to group your data? Where do you start?

How can you use the [INT÷] key to help you group and record the data?

Creación de gráficos con dibujos (continuación)

Introducción (continuación)

5. Luego, solicite a los alumnos que utilicen la tecla [Cons] para experimentar el agrupamiento de los datos.

    Ejemplo: Para explorar con grupos de 5, pida a los alumnos que:

    a. Entren [+] **5** [Cons] **0** para preparar la calculadora para contar de a 5.
    b. Entren [Cons] para contar el primer grupo de 5. Se visualiza el contador **1** y el resultado **5**.
    c. Apilen cinco fichas o cubos para representar el primer dibujo en esa categoría del gráfico.
    d. Presionen [Cons] nuevamente para contar el segundo grupo de 5. Aparecen en el visor de la calculadora el contador **2** y el resultado **10**.
    e. Apilen un segundo grupo de cinco fichas o cubos para represenar el segundo dibujo en esa categoría del gráfico.
    f. Continúen el proceso hasta que los datos de esa categoría se hayan completado (se emplearon todas las fichas o los cubos).

6. Pida a los alumnos que registren sus datos en la plantilla del gráfico de dibujos de la hoja de registro utilizando dibujos para representar el tamaño del grupo que seleccionaron.

Cómo reunir y organizar los datos

Mientras los alumnos utilizan la función [Cons] y sus contadores para agrupar los datos para el gráfico con dibujos, formule preguntas tales como:

*   ¿Qué representan las fichas o los cubos que están usando?

*   ¿Cómo decidieron agrupar sus datos? ¿Por qué decidieron agruparlos de ese modo?

*   ¿De qué modo están registrando estos grupos en el gráfico de dibujos?

*   ¿Qué harán si no tienen suficientes contadores disponibles para hacer otro grupo?

*   ¿Cómo van a registrar los restos en el gráfico de dibujos?

*   ¿Qué patrones, si los hay, encuentran cuando agrupan sus datos?

⊞ ¿Cómo pueden utilizar la calculadora para agrupar los datos?

⊞ ¿Cómo pueden utilizar la tecla [Cons] del Math Explorer para agrupar los datos? ¿Cómo pueden determinar qué entrar antes de oprimir [Cons]?

⊞ ¿Cómo pueden usar las funciones de suma con [Cons] para agrupar sus datos? ¿Dónde comienzan?

⊞ ¿Cómo pueden utilizar las operaciones de resta con [Cons] para agrupar los datos? ¿Dónde comienzan?

⊞ ¿Cómo pueden utilizar la tecla [INT÷] para facilitar la agrupación y el registro de datos?

Making Picture Graphs (continued)

Analyzing Data and Drawing Conclusions

After students have made their graphs, have them discuss the graphs as a whole group. Ask questions such as:

- What do the pictures on the graph tell you?

- What kinds of decisions did your group have to make when designing the picture graph?

- What did your group decide to do with the leftover data after it was grouped? Why did your group decide that?

- What is a question that could be answered by this graph?

- What is a question that could not be answered by this graph?

- Why do you think the designers of this graph chose each of their pictures to stand for _____ number of _____?

- What are the advantages of having a picture represent more than one piece of data?

- What are the disadvantages of having a picture represent more than one piece of data?

How did you use the [Cons] key to make the graph?

How could you use the [INT÷] key to make the graph?

How did the colored chips or cubes connect to the display you saw on your calculator?

How did the data you collected connect to the displays you saw on your calculator?

Continuing the Investigation

Have students:

- Change the number of people each picture represents to see the differences in the two graphs. Ask: Will you have more, fewer, or about the same number of pictures? How did the change affect your leftover answers from the first graph?

- Work with the other groups who chose the same category to pool all their data and make a picture graph.

- Brainstorm ways to use the information in their picture graphs, and, if possible, implement these ideas for use.

  Example: Compose a letter to the food services director presenting data about students' favorite foods.

Creación de gráficos con dibujos (continuación)

Cómo analizar los datos y sacar conclusiones

Una vez que los alumnos hayan realizado sus gráficos, pídales que discutan los mismos todos juntos formulándoles preguntas tales como las siguientes:

- ¿Qué indican los dibujos del gráfico?

- ¿Qué tipo de decisiones debió adoptar el grupo cuando diseñó el gráfico de dibujos?

- ¿Qué decidió hacer el grupo con los datos restantes después de agrupar la información? ¿Por qué el grupo tomó esa decisión?

- ¿A qué pregunta se puede responder mediante este gráfico?

- ¿A qué pregunta no se puede responder mediante este gráfico?

- ¿Por qué creen que las personas que diseñaron este gráfico eligieron cada uno de los dibujos para representar _____ veces _____?

- ¿Cuáles son las ventajas de que un dibujo represente más de un dato?

- ¿Cuáles son las desventajas de que un dibujo represente más de un dato?

Cómo continuar la investigación

Indique a sus alumnos que:

- Cambien el número de personas que representa cada dibujo para confirmar las diferencias entre los dos gráficos. Pregunte: ¿Se obtendrán más, menos o aproximadamente el mismo número de dibujos? ¿De qué modo afectó el cambio a las respuestas que quedaban del primer gráfico?

- Trabaje con los otros grupos que eligieron la misma categoría para agrupar todos sus datos y hacer un gráfico de dibujos.

- Busquen en conjunto maneras de usar la información de sus gráficos de dibujos y, si resulta posible, implementen el uso de estas ideas.

  Ejemplo: Escriban una carta para el director de los servicios de alimentación, presentando datos acerca de las comidas favoritas de los alumnos.

- 🖩 ¿Cómo usaron la tecla Cons para crear el gráfico?

- 🖩 ¿Cómo pueden usar la tecla INT÷ para crear el gráfico?

- 🖩 ¿De qué modo se relacionaron las fichas de color o los cubos con la información que aparecía en el visor de la calculadora?

- 🖩 ¿De qué modo se relacionaron los datos reunidos con la información que aparecía en el visor de la calculadora?

Name:

Making Picture Graphs
Recording Sheet

Collecting and Organizing Data

Question:	Category	Number of People
	a	
Possible answers (categories):	b	
a.		
b.		
c.	c	

Graph Title: _____

Each _____(picture) represents _____ pieces of data.

Questions we thought of while we were doing this activity:

Nombre:

Creación de gráficos con dibujos
Hoja de registro

Cómo reunir y organizar los datos

Pregunta:	

Respuestas posibles (categorías):

a.

b.

c.

Categoría	Número de personas
a	
b	
c	

Título del gráfico:_____

Cada _____(dibujo) representa _____ datos.

Preguntas que se nos ocurrieron mientras realizábamos esta actividad:

Place-Value Spinners

Math Concepts

- whole numbers
- place value
- addition
- fractions
- area
- sample space
- probability

Materials

- TI-108, Math Mate™, Math Explorer™
- **Place-Value Spinners** recording sheets
- small paper clips
- pencils
- centimeter cubes

Overview

Students will explore probability and patterns in place value by spinning two spinners and recording and analyzing the results.

Introduction

1. Have students discuss Spinner A on their recording sheets. Ask questions such as: What do you think will happen when you spin this spinner? Why do you think that? If we all spin Spinner A once and display our results in a bar graph, what do you think it would look like?

2. Make a simple spinning device by bending out one end of a paper clip. Then place the tip of your pencil through the curve at the end of the paper clip and onto the center of the spinner circle provided on page 56. Tap the paper clip with your finger to spin it about the pencil point.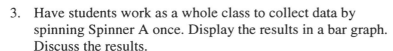

3. Have students work as a whole class to collect data by spinning Spinner A once. Display the results in a bar graph. Discuss the results.

4. Have students go through the same process to collect data for spinning Spinner B once. Discuss the results.

5. Have students work in pairs to play a game.

   Tell students: You are going to play a game with your partner where one of you will spin Spinner A and the other will spin Spinner B. Before you begin, each of you will put a marker (centimeter cube) on a number on the number line on your recording sheet. You must choose different numbers. After both of you spin, one partner will add the numbers together and the other will record the sum on the recording sheet. If your marker is on the sum of the two spins, you win a point. The game ends when one person wins 10 points.

Giradores de valor de posición

Conceptos matemáticos

- números enteros
- valores de posición
- suma
- fracciones
- áreas
- espacio muestreo
- probabilidad

Materiales

- TI-108, Math Mate™, Math Explorer™
- hojas de registro de **Giradores de valor de posición**
- clips para papel pequeños
- lápices
- cubos de 1 cm de lado

Resumen

Los alumnos explorarán la probabilidad y los patrones en valores de posición, girando dos giradores, y registrando y analizando los resultados.

Introducción

1. Pida a los alumnos que debatan acerca del Girador A de la hoja de registro. Formule preguntas tales como: ¿Qué creen que sucederá cuando giren este girador? ¿Por qué? Si todos giramos el girador A una vez y representamos los resultados en un gráfico de barras, ¿cuál será su aspecto?

2. Cree un dispositivo giratorio simple, abriendo el extremo de un clip para papel. Luego pase la punta de un lápiz por la curva del extremo del clip para papel y perfore el centro del dispositivo giratorio que se ofrece en la página 56. Golpee el clip para papel con un dedo para que gire en torno a la punta del lápiz.

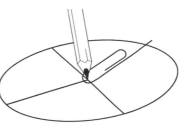

3. Pida a los alumnos que trabajen todos juntos para reunir los datos de una rotación del girador A. Represente los resultados en un gráfico de barras. Debata los resultados.

4. Solicite a los estudiantes que repitan el mismo proceso de reunión de datos de giro del girador B una vez. Debata los resultados.

5. Pida a los alumnos que trabajen de a dos para jugar un juego.

   Explique a los estudiantes: Ustedes van a jugar un juego con su compañero, en el cual uno de ustedes girará el girador A y el otro girará el girador B. Antes de comenzar, cada uno pondrá una marca (cubo de un centímetro de lado) en un número de la línea de números de la hoja de registro. Deben elegir números diferentes. Cuando los dos participantes hayan girado el dispositivo, uno de ustedes sumará los números y el otro anotará la suma en la hoja de registro. Si su marcador está situado en la suma de los dos valores de giro, el alumno se anota un punto a su favor. El juego finaliza cuando un participante gana 10 puntos.

Place-Value Spinners (continued)

Introduction (continued)

6. Have students play the game and:

   a. Collect data about their spins.
   b. Record the data on their recording sheets.
   c. Analyze the data in order to develop better strategies for winning the game.

   Note: Encourage students to use a calculator for finding sums.

Collecting and Organizing Data

While students are playing the game and collecting data about the sums of the spins, ask questions such as:

- What numbers are you putting your counter on? Why are you choosing those numbers?

- Is every number a possible winner? Why or why not?

- Which numbers do you think are more likely to win than others? Why do you think that? Does the data you are collecting seem to support your ideas?

- What strategies are you using to play the game? Why did you choose those strategies?

- What patterns do you see in the sums you are spinning? How are those patterns useful?

How can you use the calculator to help you find the sums?

What patterns do you see that will help you find the sums without using the calculator?

Analyzing Data and Drawing Conclusions

After students have played the game and collected their data, have them discuss the results as a whole group. Ask questions such as:

- What sums are possible outcomes?

- How do you know that is all of them?

- Are all the sums equally likely, or do some happen more often than others? Why?

- What strategies did you use to play the game? Why did you choose those strategies?

How did you use the calculator to help you in this game?

Were you able to stop using the calculator? Why?

When are calculators most useful?

When are calculators not very useful?

Giradores de valor de posición (continuación)

Introducción (continuación)

6. Luego, pida a los alumnos que jueguen y:

   a. Reúnan datos sobre los giros.
   b. Registren los datos en la hoja de registro.
   c. Analicen los datos para desarrollar mejores estrategias para ganar el juego.

   Nota: Anime a los alumnos a usar una calculadora cuando deban resolver sumas.

Cómo reunir y organizar los datos

Mientras los alumnos juegan y reúnen datos sobre las sumas de los giros, formule preguntas tales como las siguientes:

- ¿En qué número están usando el contador? ¿Por qué eligen esos números?

- ¿Tienen todos los números las mismas probabilidades de ganar? ¿Por qué?

- ¿Qué números creen que tienen más probabilidades de ganar? ¿Por qué piensan eso? ¿Parecen respaldar esta opinión los datos reunidos?

- ¿Qué estrategias están usando para el juego? ¿Por qué eligieron esas estrategias?

- ¿Qué patrones observan en las sumas? ¿Son esos patrones útiles?

🖩 ¿Cómo pueden utilizar la calculadora para encontrar las sumas?

🖩 ¿Qué patrones observan que les permitirán encontrar las sumas sin usar la calculadora?

Cómo analizar los datos y sacar conclusiones

Una vez que los alumnos hayan completado el juego y reunido sus datos, pídales que debatan acerca de los resultados en grupo formulándoles preguntas tales como:

- ¿Qué sumas son resultados posibles?

- ¿Cómo sabe que esas son todas?

- ¿Tienen todas las sumas la misma probabilidad? ¿Parecen algunas más frecuentes que otras? ¿Por qué?

- ¿Qué estrategias utilizaron en el juego? ¿Por qué eligió esas estrategias?

🖩 ¿Cómo usaron la calculadora para facilitar el juego?

🖩 ¿Pudieron dejar de usar la calculadora? ¿Por qué?

🖩 ¿En qué momento resultan más útiles las calculadoras?

🖩 ¿En qué momento no resultan tan útiles las calculadoras?

Place-Value Spinners (continued)

Analyzing Data and Drawing Conclusions (continued)

- What if you changed the sizes of the sections on the spinners? How do you think it would change the game?

- What if you changed the numbers on the spinners? How do you think it would change the game?

- What are the advantages, if any, of knowing when something is more likely to happen than something else?

- What are the disadvantages, if any?

Continuing the Investigation

Have students:

- Change the sizes of the sections on the spinners, predict how the likelihood of the outcomes will change, and collect data to compare to their predictions.

- Change the numbers on the spinners, predict how the outcomes will change, and collect data to compare to their predictions.

1995 Texas Instruments Incorporated. ™ Trademark of Texas Instruments Incorporated.

Giradores de valor de posición (continuación)

Cómo analizar los datos y sacar conclusiones (continuación)

- ¿Qué sucedería si cambiaran el tamaño de las secciones de los giradores? ¿De qué modo creen que cambiaría el juego?

- ¿Qué sucedería si cambiaran los números de los giradores? ¿De qué modo creen que esto cambiaría el juego?

- ¿Cuáles son las ventajas, si las hay, de conocer cuándo es más probable que suceda una cosa que otra?

- ¿Cuáles son las desventajas, si las hay?

Cómo continuar la investigación

Indique a sus alumnos que:

- Cambien el tamaño de las secciones de los giradores, predigan las probabilidades de que cambien los resultados y reúnan datos para comparar con sus predicciones.

- Cambien los números en los giradores, predigan de qué modo cambiarán los resultados y reúnan datos para comparar con sus predicciones.

Name:

Place-Value Spinners
Recording Sheet

Collecting and Organizing Data

Number Line

11 12 13 14 15 16 17 18 19 20 21 22 23 24 25 26 27 28 29 30 31 32 33

A:

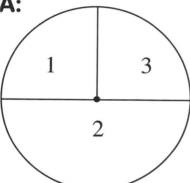

Sum of Two Spins	Number of Times It Happened

B:

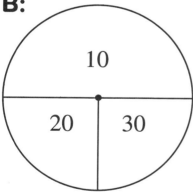

Scores:

Player 1: _____

Player 2: _____

Patterns we found while we were doing this activity:

Nombre:

Giradores de valor de posición
Hoja de registro

Cómo reunir y organizar los datos

11 12 13 14 15 16 17 18 19 20 21 22 23 24 25 26 27 28 29 30 31 32 33

A:

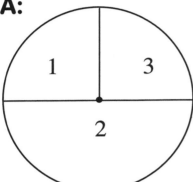

Suma de dos giros	Número de veces que sucedió

B:

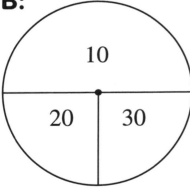

Puntuación:

Participante 1: _____

Participante 2: _____

Patrones que encontramos mientras realizábamos esta actividad:

Bibliography for Links to Literature

Adams, P., *There Was an Old Lady Who Swallowed a Fly*. Ladybird Books, 1990.

Adams, P., *There Were Ten in a Bed*. Child's Play, 1979.

Aker, S., *What Comes in 2's, 3's, and 4's?* Simon and Shuster, 1990.

Anno, Mitsumasa, *Anno's Counting House*. Putnam G.P. & Sons, 1982.

Carle, E., *Rooster's Off to See the World*. Picture Book Studio, 1972.

Crews, D., *Ten Black Dots*. Greenwillow Books, 1986.

Ernst, L., *Sam Johnson and the Blue Ribbon Quilt*. Lothrop, 1983.

Flournoy, V., *The Patchwork Quilt*. Dial Books for Young Readers, 1985.

Giganti, Jr., P., *Each Orange Had 8 Slices. A Counting Book*. Greenwillow Books, 1992.

Hopkinson, D., *Sweet Clara and the Freedom Quilt*. Alfred A. Knopf, 1993.

Hutchins, P., *1 Hunter*. Morrow, 1982.

Mayer, M., *These Are My Pets*. Western Publishing Company, Inc., 1988.

Mayer, M., *When I Get Bigger*. Western Publishing Company, Inc., 1985.

Morozumi, A., *One Gorilla*. FS&G, 1993.

O'Keefe, S.H., *One Hungry Monster*. Little Brown & Co., 1989.

Polacco, P., *The Keeping Quilt*. Simon and Schuster Books for Young Readers, 1988.

Raffi, *Five Little Ducks*. Crown Books for Young Readers, 1992.

Activity Content Index

Activity Name	Page Number	Patterns	Whole Numbers	Fractions	Place Value	Operations	Estimation	Measurement	Geometry	Probability Statistics
Action-Packed Stories	2	✓	✓		✓	+ −				
Patterns in Counting	6	✓	C		✓	+				
It's the Place that Counts	9	✓	C		✓	+				
a-MAZE-ing Secret Paths	14	✓	C			+ −	✓			
Action-Packed Addition Patterns	19	✓	✓			+				
Action-Packed Subtraction Patterns	23	✓	✓			−				
Predictable Patterns with Addition	27	✓	C			+ ×				
Design a Quilt	32	✓	✓			+			P	
Double Your Design	36	✓	✓			+			S	
Symmetry with Pattern Blocks	40	✓	✓			+			S	
Reading Picture Graphs	45	✓	✓			+ ×				G
Making Picture Graphs	49	✓	✓			+ ×				G
Place-Value Spinners	53	✓	✓	✓	✓	+		A	✓	

C = Comparing whole numbers

A = Area

P = Polygons

S = Symmetry

G = Graphing